Terrorism in Cyberspace

Terrorism in Cyberspace

The Next Generation

Gabriel Weimann

Woodrow Wilson Center Press
Washington, D.C.

Columbia University Press
New York

EDITORIAL OFFICES

Woodrow Wilson Center Press
Woodrow Wilson International Center for Scholars
One Woodrow Wilson Plaza
1300 Pennsylvania Avenue NW
Washington, DC 20004-3027
www.wilsoncenter.org

ORDER FROM

Columbia University Press
Sales Department
61 West 62nd Street
New York, NY 10023
Telephone: 212-459-0600 ext. 7129
http://cup.columbia.edu

© 2015 by Gabriel Weimann

2 4 6 8 9 7 5 3 1

Library of Congress Cataloging-in-Publication Data

Weimann, Gabriel, 1950–
   Terrorism in cyberspace : the next generation / Gabriel Weimann.
   pages cm
   ISBN 978-0-231-70448-9 (cloth)
   ISBN 978-0-231-70449-6 (paper)
   ISBN 978-0-231-80136-2 (ebook)
   1. Cyberterrorism. 2. Terrorism. I. Title.
   HV6773.15.C97 W45 2014
   363.325—dc23
        2014042761

Cover design: Naylor Design, Inc.

References to websites (URLs) were accurate at the time of writing.
Neither the author nor Woodrow Wilson Center Press and Columbia
University Press is responsible for URLs that may have expired or
changed since the manuscript was printed.

Wilson Center

Woodrow Wilson International Center for Scholars

The Wilson Center, chartered by Congress as the official memorial to President Woodrow Wilson, is the nation's key nonpartisan policy forum for tackling global issues through independent research and open dialogue to inform actionable ideas for Congress, the Administration, and the broader policy community.

Conclusions or opinions expressed in Center publications and programs are those of the authors and speakers and do not necessarily reflect the views of the Center staff, fellows, trustees, advisory groups, or any individuals or organizations that provide financial support to the Center.

Please visit us online at www.wilsoncenter.org.

Jane Harman, Director, President, and CEO

Contents

PART III.   Future Threats and Challenges

Figures

ix

Foreword

Bruce Hoffman

"The story of the presence of terrorist groups in cyberspace has barely begun to be told," Gabriel Weimann reflected almost a decade ago in his seminal work, *Terror on the Internet*. Even so accomplished a scholar of communications as Professor Weimann, however, could not have anticipated the changes and advances in technology that would revolutionize terrorism during the second decade of the twenty-first century.

Much like Afghanistan in the 1990s, places like Syria and Iraq today have often been described as the "perfect jihadi storm": magnets for foreign fighters, where violence is theologically justified by clerics issuing fatwas (religious edicts) and where rebels—including core al-Qaeda loyalists like Jabhat al-Nusra (the Al-Nusra Front) and renegade groups such as the Islamic State of Iraq and al-Sham (ISIS)—benefit from the largesse of wealthy Arabian Gulf patrons. But a critical distinction between the struggle in Afghanistan during the closing decades of the twentieth century and in Syria and Iraq in the early twenty-first-century is the evolution of information technology and communications that has unfolded since *Terror on the Internet* was published in 2006. The growth and communicative power of social networking platforms such as Facebook, YouTube, Twitter, Instagram, Flickr, and WhatsApp have transformed terrorism: facilitating both ubiquitous and real-time communication between like-minded radicals with would-be recruits and potential benefactors, thus fueling and expanding the fighting and bloodshed to a hitherto almost unprecedented extent.

As *Terrorism in Cyberspace: The Next Generation* so ably explains, it is not uncommon nowadays for foreign fighters prosecuting these conflicts to amass thousands of followers on platforms such as Twitter and Facebook. They communicate with their audiences often on a daily basis—and sometimes multiple times each day—providing first-hand, immediate accounts of heroic battles and more mundane daily activities, making jihad accessible and comprehensible on a uniquely intimate and personal basis. Fighters invite, motivate, animate, and summon their Twitter followers and Facebook friends to travel to Syria and Iraq and partake of the holy war against the Assad and Maliki regimes. Blatant sectarian messaging and divinely ordained clarion calls to resist Persian domination and help determine the outcome of the eternal struggle between Sunni and Shi'a—and the latter's Alawite satraps—provide additional, compelling incentives. Indeed, a recent ISIS recruitment video posted on the Internet featured heavily armed militants with distinctive British and Australian accents trumpeting the virtues of jihad and the ineluctable religious imperative of joining the caravan of martyrs. It is therefore not surprising to find that all of al-Qaeda's most important affiliates—al Shabaab, Ansar al-Sharia, Boko Haram, the Abdullah Azzam Brigades, Al-Qaeda in the Islamic Maghreb, Al-Qaeda in the Arabian Peninsula, the Al-Nusra Front, and the Afghan Taliban—as well as the outlawed ISIS, all have Twitter accounts on which they regularly tweet.

According to Weimann, social media provides manifold advantages to terrorists. "New communication technologies," he explains, "such as comparatively inexpensive and accessible mobile and web-based networks, create highly interactive platforms through which individuals and communities share, co-create, discuss, and modify content." Interactivity, reach, frequency, usability, immediacy, and permanence are the benefits reaped by terrorist groups exploiting and harnessing these new technologies.

Much as *Terror on the Internet* filled a conspicuous gap in the literature on terrorists and terrorism when it was first published, *Terrorism in Cyberspace* does the same now. It represents the next step in its author's decades-long quest to map, analyze, and understand the evolution of terrorist communications that has occurred since the advent of the Internet and this new form of mass communication. When Weimann first began to examine this phenomenon in 1998, he recounts, there were perhaps no more than a dozen terrorist groups online—including al-Qaeda. Today, Weimann's attention is consumed by a staggering 10,000 terrorist websites, in addition to the innumerable social media platforms proliferating throughout cyberspace. "This trend," Weimann warns, "is combined with the emergence of lone wolf terrorism: attacks by individual terrorists who are not members of any terrorist organization." He describes how lone wolf terrorism is the "fastest-growing kind of terrorism, especially in the West, where all recent lone wolf attacks involved individuals who were radicalized, recruited, trained, and even launched on social media platforms." The implications for law enforcement and intelligence and security agencies, already stretched thin by splintering groups, multiplying threats, and their own diminished budgets and resources, are fundamentally disquieting.

Weimann believes that government counterterrorism efforts must adjust and recalibrate existing strategies and tactics to meet the immense challenges presented by these new communications and propaganda platforms. The considerable knowledge and experience that communications experts in the United States have acquired in running political and advertising campaigns, he argues, need to be appropriated and redirected to countering terrorism and terrorist use of the Internet and social media. To do so, Weimann contends, we need to better anticipate future trends in terrorist communications and better prepare to counter them before they actually materialize.

Terrorism in Cyberspace embodies the hallmarks of Weimann's decades of scholarship: presenting a comprehensive, thoughtful, and sober analysis—supported by voluminous empirical evidence and trenchant, revealing examples. Years from now, when historians seek to explain how the threat from al-Qaeda and associated groups as well as still more radical offshoots surfaced and multiplied throughout 2013 and 2014, *Terrorism in Cyberspace* will be indispensable in revealing how all this came to pass. For that reason, among others, this book is essential reading for anyone seeking to understand the dynamics of contemporary terrorism and its exploitation of modern media technology.

Bruce Hoffman
Washington, D.C.
June 2014

Acknowledgments

Terror in Cyberspace took me 16 years of research, analysis, and writing. These years have been spent in the darkest alleys of the Internet, in the most gruesome and violent terrorist sites and in the scariest corners of the cyber world. This experience is far from being civilized, humane, or comforting. And yet even in this dark project, there were many lights, many reassuring experiences of support, compassion, and friendship. These came from a group of people who contributed, in various ways, to my work and provided the kind counterbalance to the often frightening worlds I have studied and experienced.

First and foremost, I would like to thank my wife Nava for standing beside me throughout my career and while working on this book. She has been my inspiration and motivation, she is my solid rock, and I dedicate this book to her. I also thank my children, Oren and Dana: hopefully, they will read this book and understand why I spent so much time in front of my computer or traveling so frequently and so far.

The research reported here could not have been done without the contributions of many individuals and institutions all over the world. Several grants and foundations contributed directly and indirectly to this project, but the most important hosting and funding organization was the Woodrow

Wilson International Center for Scholars in Washington, D.C. The Wilson Center and its devoted team were a home and a family for me during my year as a Fellow at the Center and long after that. Many of the Center's excellent staff deserve my gratitude, but I would like to single out Robert S. Litwak, the vice president for scholars and director of international security studies at the Center. Without him, this book would never have seen the light: Rob was the one who introduced the Center to me; supported my fellowship proposal; and during my time at the Center was my counselor, a source of advice and guidance, a spiritual and intellectual mentor. Joseph Brinley, the director of Woodrow Wilson Center Press, was always there to support and advise and finally to advance the publication of this book.

At the Wilson Center, I was assisted by many devoted staff members, and I can list only a few: The energetic and resourceful Jane Harman, the Center's director, president, and CEO; Center executive vice president Andrew Selee; Middle East Program director Haleh Esfandiari; fellowship specialist Kimberly Conner; coordinator of fellows' services Arlyn Charles; Lindsay Collins; Lea Shanley; Meg King; Domenick Jervis; the librarians and the great computer technicians who were often hassled by my inquiries and computer problems. This research project was carried out with the help of numerous research assistants. Two of them who worked with me at the Wilson Center deserve my appreciation for their contribution and involvement: Jannis Jost, a promising young and very skillful researcher from Germany, and the talented Daniela Sainz from the United States.

Several institutions and organizations assisted my data collection and translation of websites. They include the Pew Internet & American Life Project, the Middle East Media Research Institute (MEMRI), the SITE Intelligence Group headed by Rita Katz, and the Internet Haganah. Many of my colleagues and friends are to be thanked for their support, advice, and encouragement. I will name only a few: Bruce Hoffman from Georgetown University, who was always a friend as well as a model of a top researcher on terrorism; Elliott Milstein; Eugene Rothman; Dorothy Denning; Yoram Peri; Pnina Peri; Louis Goodman; Boaz Ganor; Joshua Sinai; Hanna and Arie Kruglanski; Nava and Eitan Tadmor; Josephine and Rami Levi; Gail and Yash Shirazi; Ofer Goren; Zion Avissar and many others. I wish to express my appreciation and gratitude to two reviewers for their excellent suggestions and comments, and to Shannon Granville, my text editor: I have worked with several editors on my past books and publications, but Shannon was certainly the most devoted one. Her in-depth editing improved the text significantly, and I am more than fortunate to have had her as an editor.

This project, despite its relevance for counterterrorism, was not supported by any security, military, or federal agency. Yet, in my presentations and lectures, I have noted the presence of representatives from such organizations, agencies, and units. I hope they did learn from my work about this new arena—the cyberspace—as used by modern terrorism, but also about the need to consider the limits and boundaries that democracies should apply to their war on terrorism in the virtual arena.

Terrorism in Cyberspace

Introduction

Can we declare the war on terrorism to be over? According to a growing body of opinion in Washington and elsewhere, we should declare the end of the war on terrorism—or at least recognize that the topic has been grievously overhyped and exaggerated. John Mueller of Ohio State University and Mark G. Stewart of the University of Newcastle in Australia coauthored a paper looking at the costs and benefits of counterterrorism spending (Mueller and Stewart 2011). They argue that, based on the US government's spending on counterterrorism, we are grossly overestimating the risks of terrorism. In the wake of the 2011 Arab Spring, politicians, government officials, and journalists expressed an optimistic view regarding the future of global terrorism. Thus, for example, a senior official in the State Department told the *National Journal*: "The war on terror is over: Now that we have killed most of al Qaida, now that people have come to see legitimate means of expression, people who once might have gone into al Qaida see an opportunity for a legitimate Islamism" (cited in Hirsh 2012). In a similar vein, CNN national security analyst Peter Bergen declared in October 2011 that after the death of Osama bin Laden, the war on terror was over (Crabbe 2011).

However, the National Counterterrorism Center's 2011 annual report (released in June 2012) reveals that terrorism is far from dying out. In one year, there were more than 10,000 attacks classified by the US government as terrorism; these attacks claimed a total of 12,500 lives worldwide. The report described an increase in radicalization worldwide and an enmity toward the United States around the globe (National Counterterrorism Center 2012). This has led terrorism experts like Bruce Hoffman to conclude that the war on terrorism may have been very effective tactically—in killing and capturing terrorists—but it is not having a positive effect in changing the environment that promotes or that gives rise to terrorism: "Currently, al-Qaeda continues to maintain extensive operational environments and safe havens in Pakistan, Somalia, and Yemen. Additionally, its presence in Syria has become commonplace and will be pivotal in the organization's future" (Hoffman et al. 2012). Others, like Mark Katz from George Mason University, even have argued, "One prediction about the 'War on Terror' can be made with great confidence: It is not going to end any time soon, or even dramatically subside. There are several possible ways, though, in which it could evolve" (Katz 2011, 1).

In 2013, the Bipartisan Policy Center released a new terrorism threat assessment that warns that individuals who have self-radicalized over the Internet pose the most imminent threat to US homeland security, even as al-Qaeda has managed to spread its jihadist ideology across a larger geographical area than ever before. The report, *Jihadist Terrorism: A Threat Assessment*, provides a comprehensive review of al-Qaeda and its affiliates, and provides legislative and executive recommendations on how best to improve the US counterterrorism and homeland security strategy (Bergen et al. 2013). The report was authored by several members of the center's Homeland Security Project, which is led by former 9/11 Commission co-chairs Tom Kean and Lee Hamilton. "Today, the United States faces a different terrorist threat than it did on 9/11 or even three years ago," states the report's executive summary. "As a result, many counterterrorism officials believe the chances of a large-scale, catastrophic terrorist attack by al-Qaeda or an al-Qaeda-affiliated or -inspired organization occurring in the United States are small." But while the core of al-Qaeda may be in decline, "'al Qaeda-ism,' the movement's ideology, continues to resonate and attract new adherents" (Bergen et al. 2013, 5). One example of the emergence of new jihadi groups is the case of the Islamic State (IS), also previously known as the Islamic State of Iraq and the Levant (ISIL) and the Islamic State of Iraq and Syria (or al-Sham) (ISIS). IS claims religious authority over all Muslims worldwide and aspires to exercise direct political control over many Muslim-inhabited regions. The group's initial aim was to establish a caliphate in the Sunni-majority regions of Iraq, but following its involvement in the Syrian Civil War this goal expanded to include control over the Sunni-majority areas of Syria. On June 29, 2014, it proclaimed that it had established a caliphate and named the Al-Qaeda in Iraq leader Abu Bakr al-Baghdadi as its caliph; the group was renamed the Islamic State (Withnall 2014). The United Nations Security Council, the United Kingdom, the United States, and other states have officially designated the group as a foreign terrorist organization (United Nations Security Council 2014, Home Office 2014, Department of State 2014a).

More than a decade after September 11, 2001—after witnessing the Arab Spring and the death of Osama bin Laden in 2011—is it finally time to end the War on Terror? This should be stated as an empirical question, a scholarly challenge to be answered by data, analysis, and findings rather than by suppositions and speculations. Moreover, there is a need to recognize that terrorism has changed, that the nature of terrorist threats has changed, and that future trends may bring about new forms of terrorism and threats. This is where the project on monitoring terrorist presence on the Internet, and its numerous online platforms, becomes so important and useful. Studying terrorist communication online is one critical means of early warning or scanning of the horizon for potential future threats, as well as a method of keeping on top of evolving trends in terrorism.

The fact that cyberspace has became an important, if not the most important, arena for terrorist communications (in addition to a potential battlefield) is no longer questioned.[1] There are, however, uncertainties and doubts about the future directions of online terrorism. As William McCants asserted during his testimony on December 2011 before the House Homeland Security Subcommittee on Counterterrorism and Intelligence,

> There is little research to go on, which is striking given how data-rich the Internet is. In hard numbers, how widely distributed was Zawahiri's last

---

[1] The Department of Defense defines cyberspace as a global domain within the information environment consisting of the interdependent network of information technology infrastructures and associated data, including the Internet, telecommunications networks, computer systems, and embedded processors and controllers. See Deputy Secretary of Defense Memorandum, Subject: *The Definition of Cyberspace*, May 12, 2008.

message? Did it resonate more in one U.S. city than another? Who were its main distributors on Facebook and YouTube? How are they connected with one another? This sort of baseline quantitative research barely exists at the moment. (McCants 2011)

This book attempts to fill this research gap by answering the following three research questions:

- What are the new faces of online terrorism?
- What can be expected in the near future?
- How can we counter these trends?

Research Question 1: What are the new faces of online terrorism?

"[Terrorists'] online activities offer a window onto their methods, ideas, and plans."

—Evan Kohlmann, "The Real Online Terrorist Threat" (2006)

The typical loosely knit network of cells, divisions, and subgroups of modern terrorist organizations finds the Internet both ideal and vital for inter- and intragroup networking. Websites, however, are only one of the Internet's services to be hijacked by terrorists: other facilities include email, chatrooms, e-groups, forums, virtual message boards, YouTube, and Google Earth. The rise of "internetted" terrorist groups is part of a broader shift to what John Arquilla and David Ronfeldt have called "netwar" (Arquilla and Ronfeldt 2001, 2003; Arquilla, Ronfeldt, and Zanini 2001). Netwar refers to an emerging mode of conflict and crime at societal levels, which involves measures short of traditional war in which the protagonists are likely to consist of small, dispersed groups who communicate, coordinate, and conduct their campaigns in an "internetted" manner, and without a precise central command. Today, all terrorist organizations, large or small, have their own websites, Facebook pages, or uploaded YouTube videos (Weimann 2006b, 2014a; Hoffman 2006b).

However, terrorism itself has changed, both in structure and in mode of operation. New forms of terrorism have transformed into segmented networks instead of the pyramidal hierarchies and command-and-control systems that govern traditional insurgent organizations. Take, for example, the case of "lone wolves." Lone wolf terrorism is the fastest growing kind of terrorism (Weimann 2014a, 2014b). Before 9/11, the men who went to terrorist camps and to jihadi mosques where radical imams preached jihad were seen as constituting the largest terror threat. Since 9/11, a gradual change has occurred. The real threat now comes from the single individual: the "lone wolf" living next door, being radicalized on the Internet and plotting strikes in the dark. Acts of lone wolf terrorism have been reported in Australia, Canada, Denmark, France, Germany, Italy, the Netherlands, Poland, Portugal, Russia, Sweden, the United Kingdom, and the United States. However, despite the alarming increase in lone wolf terrorism, there seems to be a gap between the perceived threat of lone wolf terrorism on the one hand and the almost exclusive scholarly focus on group-based terrorism on the other hand. The need for more conceptual and empirical examinations of lone wolf terrorism may lead, as this study suggests, to revealing the lone wolves' reliance on modern communication platforms (Weimann 2014b).

Thus, our first goal will be to summarize the findings on the shifts in terrorist use of online platforms. What are the new faces of online terrorism, and how do these new forms of online presence reflect changes in the structure and *modus operandi* of new terrorism?

Research Question 2. What can be expected in the near future?

The next generation of terrorists won't be mindless hordes of thugs living a hand-to-mouth existence in Afghanistan. The young kids that they are radicalizing today are studying mathematics, computer science and engineering. They will grow up and realize 'I'm too valuable to stuff dynamite around my waist and walk into a crowded cafe.' And they will think very differently about how they can attack their perceived enemies. The internet will be another tool in their toolbox.

—Dan Verton, *Black Ice: The Invisible
Threat of Cyber-Terrorism* (2003, 18)

One of the most difficult challenges faced by al-Qaeda today is the ongoing loss of a large part of its first-, second-, and even third-generation leadership, some of whom have been assassinated or arrested. Still others have completely dissociated themselves from the organization and its terrorist methods. As observed in a recent Europol report on terrorism trends, "As a consequence of sustained military pressure, al-Qaeda core have publicly discouraged sympathizers from travelling to conflict zones in order to join them. It has instead promoted the idea of individually planned and executed attacks in Western countries without the active assistance of any larger organization" (Europol 2012, 19).

The Europol report, published in April 2012, highlighted the importance of the Internet. The report states that the Internet has become the "principal means of communication" for extremist groups, which now have a "substantial online presence" (Europol 2012, 6). As well as its use for propaganda, recruitment, fund-raising, and planning—all facilitated by social media— the Internet has the potential to be utilized in cyberattacks on the operating systems of vital infrastructure in European Union member states. This trend of online recruitment, radicalization, and activation by terrorists may indicate the growing reliance of future terrorism on online platforms, as well as their adaption of new cyber platforms. The second goal of this project is focused on the future: what can we expect in the coming years in terms of terrorist presence on the Net and its new platforms?

One of the most alarming future scenarios is that of cyberterrorism. Cyberterrorism is commonly defined as the use of computer network devices to sabotage critical national infrastructures such as energy, transportation, or government operations. Cyberterrorism is in fact the use of cyber technology to commit terrorism. Given the range of cyberterrorism activities described in the literature, this simple definition can be expanded to include the use of cyber capabilities to conduct enabling, disruptive, and destructive militant operations in cyberspace to create and exploit fear through violence or the threat of violence in the pursuit of political change (Brickey 2012). The premise of cyberterrorism is that as modern infrastructure systems have become more dependent on computerized networks for their operation, new vulnerabilities have emerged—"a massive electronic Achilles' heel" (Lewis 2002). Cyberterrorism is an attractive option for modern terrorists who value its potential to inflict massive damage, its psychological impact, and its media appeal.

Research Question 3. How can we counter these trends?

All too often we are reminded that terrorism continues to inflict pain and suffering on people all over the world. Hardly a week goes by without an act of terrorism taking place somewhere in the world, indiscriminately affecting innocent people, who just happened to be in the wrong place at the wrong time. Countering this scourge is in the interest of all nations and the issue has been on the agenda of the United Nations for decades.

—United Nations Action to Counter Terrorism[2]

The Internet is clearly changing the landscape of political discourse and advocacy. It offers new and inexpensive methods to collect and publish information, to communicate and coordinate action on a global scale, and to reach out to world public opinion as well as decision makers. The Internet benefits individuals and small groups with few resources as well as large or well-funded organizations. It facilitates activities such as educating the public and the media, raising money, forming coalitions across geographical boundaries, distributing petitions and action alerts, and planning and coordinating regional or international events. It allows activists in politically repressive states to evade government censors and monitors. Thus, the Internet could have become a peaceful and fruitful forum for the resolution of conflicts, and yet it has also become a useful instrument for terrorists. Their use of this liberal, free, easy-to-access medium is indeed frightening. Nonetheless, one should consider that the fear that terrorism inflicts can be and has been manipulated by politicians to pass questionable legislation that undermines individual rights and liberties—legislation that otherwise would not stand a chance of being accepted by the public. It is important to assess the real threat posed by terrorist groups using the new information technology, keeping in mind that government action against it could easily go beyond acceptable limits (Weimann 2007b, 214).

Fighting terrorism raises the issue of countermeasures and their prices: "Terrorist tactics focus attention on the importance of information and communications for the functioning of democratic institutions; debates about how terrorist threats undermine democratic practices may revolve around freedom of information issues" (Arquilla and Ronfeldt 2001, 14). Responding to terrorism in the Internet is an extremely sensitive and delicate issue, since most of the rhetoric disseminated on the Internet is considered protected speech under the United States' First Amendment and similar provisions in other societies. Since the advent of the Internet, the US Central Intelligence Agency, the National Security Agency, the Federal Bureau of Investigation, and intelligence and security services all over the world have seen it as both a threat and a tool. There are numerous efforts, some of which are kept secret and some of which are not, to apply systems, measures, and

---

[2] See the United Nations Action to Counter Terrorism website at www.un.org/en/terrorism/.

defense mechanisms against terrorists on the Internet. Besides the legal and practical issues, Internet counterterrorism suffers from a lack of strategic thinking. Various measures have been suggested, applied, replaced, changed, and debated, yet there was never an attempt to propose a general model of online counterterrorism strategy. Countering terrorist usage of the Internet to further ideological agendas will require a strategic, government-wide (interagency) approach to designing and implementing policies to win the war of ideas. This book suggests the notion of "noise" in communication theory as a basic theoretical framework to conceptualize various measures and their applicability (Von Knop and Weimann 2008). The last part of this book will not only review the various countermeasures applicable to the case of online terrorism but also examine their "prices" in terms of civil liberties.

The Terror on the Internet Project

This project is based on a database collected during 15 years of monitoring thousands of terrorist websites (Weimann 2006b, 2007a, 2008a, 2010c, 2012b). The population for this study was defined as the Internet sites of terrorist movements as they have appeared and will appear in the period between January 1998 and October 2013. To determine the population of terrorist organizations, the study uses the US Department of State's list of foreign terrorist organizations (2014a). This raises the sensitive issue of defining terrorism. To study terrorism, whether on the Internet or elsewhere, it is first necessary to define just what constitutes a terrorist organization.

Although most people know terrorism when they see it, academics and scholars are unable to agree on a precise definition. "Terrorism" may well be the most important word in the political vocabulary these days, as was remarked by one of its most prominent students, Alex P. Schmid, director of the Centre for Study of Terrorism and Political Violence at the University of St. Andrews (Schmid 2004). Nevertheless, little has changed since 1984, when Schmid first concluded that even though "[academic researchers from many fields] have spilled almost as much ink as the actors of terrorism have spilled blood," they have not yet reached a consensus on what terrorism is (reprinted in Schmid and Jongman 2005, introduction). In their attempts to construct a working definition, Schmid and Albert Jongman (1988, 2005) presented the results of a survey of leading academics in the field, each of whom was asked to define terrorism. From these definitions, the authors isolated the following recurring elements, in order of their statistical appearance in the definitions: Violence or force (appeared in 83.5 percent of the definitions); political (65 percent); fear or emphasis on terror (51 percent); threats (47 percent); psychological effects and anticipated reactions (41.5 percent); discrepancy between the targets and the victims (37.5 percent); intentional, planned, systematic, organized action (32.0 percent); methods of combat, strategy, tactics (30.5 percent). From these elements, the following working definition emerged:

> Terrorism is an anxiety-inspiring method of repeated violent action, employed by (semi-) clandestine individual, group or state actors, for idiosyncratic, criminal or political reasons, whereby—in contrast to assassination—the direct targets of violence are not the main targets. The immediate human victims of violence are generally chosen randomly (targets of opportunity) or selectively (representative or symbolic targets) from a target population, and serve as message generators. Threat- and violence-based communication processes between terrorist (organization), victims (imperiled), and main targets (audience(s)) are used to manipulate the main target, turning it into a target of terror, a target of demands, or a target of attention, depending on whether intimidation, coercion, or propaganda is primarily sought." (Schmid and Jongman 2005, 28)

This definition of terrorism has been adapted in similar forms by such contributors and analysts as *Jane's Intelligence Review*, the US Department of State, and most terrorism scholars.

To locate the online terrorist sites, frequent systematic scans of the Internet were conducted using the various keywords and names of organizations in the database. First, the standard search engines (e.g., Google, Yahoo!, Bing) were used. The Internet is a dynamic arena: websites emerge and disappear, change addresses, or are reformatted. Years of monitoring the terrorist presence online has provided information on how to locate their new sites, how to search in chatrooms and forums of supporters and sympathizers for the new "addresses," and how to use links in other organizations' websites to update existing lists. This was often a Sisyphean effort, especially since in certain instances—for instance, al-Qaeda's sites—the location and the contents of the sites changed almost daily.

When this research began in the late 1990s, there were merely a dozen terrorist websites (Tsfati and Weimann 2002); by 2000, virtually all terrorist groups had established their presence on the Internet, and in 2003 there were more than 2,600 terrorist websites (Weimann 2006b). The number rose dramatically, and by October 2013, the project archive contained more than 9,600 websites serving terrorists and their supporters. The process of monitoring terrorist websites involves tracking them, downloading their contents, translating the messages (texts and graphics), and archiving them according to a preset coding system. The project enjoyed funding from various academic foundations, including the United States Institute of Peace and the National Institute of Justice.

The study involved (and will continue to involve) continuous cooperation with various organizations and institutes involved in similar studies and data collection, including the SITE Intelligence Group, which monitors jihadi websites; the Middle East Media Research Institute (MEMRI); the RAND Corporation; Internet Haganah; and the International Institute for Counter-Terrorism and its Jihadi Websites Monitoring Group. This project allowed for various content analyses, including the following studies (all published in peer-reviewed journals) whose findings have been incorporated into this text:

- Terrorist online radicalization, recruitment, and mobilization (Weimann 2005b, 2005c, 2006d, 2007a, 2007b, 2008d, 2009a): These studies examined how the Internet has been used a recruitment, radicalization, and mobilization tool, such as through the creation of virtual training camps and plans to target Western financial systems and economic infrastructure.
- Terrorists' use of online fatwas (Weimann, 2009d, 2011b): These studies revealed how the Internet is used to justify actions and solve internal debates through the posting of online fatwas (religious rulings).
- Al-Qaeda's reliance on the Internet (Weimann 2008a, 2008f, 2011a): These studies monitored the changes in al-Qaeda's uses of the Internet and highlighted the correlations between these changes and the changing structural characteristics of al-Qaeda.
- The threat of cyberterrorism (Weimann 2005a, 2006a, 2006b, 2008b): These studies examined the potential for cyberattacks based on references to such actions in terrorist chatter.
- Terrorist debates online (Weimann 2006c, 2009d): These studies investigated the issues debated by terrorists online, including types of action, legitimate targets, and the use of women and children.

- Narrowcasting: Terrorists targeting children and women online (Weimann 2008c, 2009b): These studies looked at the growing use of online platforms to target specific subpopulations, a marketing technique adapted by Internet-savvy terrorists.
- Terrorists' use of new media (e.g., YouTube, Twitter, MySpace, Facebook) (Weimann and Vail Gorder 2009, Mozes and Weimann 2010, Weimann 2010b, 2010c, 2011a, Kennedy and Weimann 2011): These studies revealed the growing use and sophistication by terrorists of social networking online and the targeted populations.
- Lone wolf terrorism and the Internet (Weimann 2012a, 2014a, 2014b): These studies addressed the growing popularity of using the Internet to attract, radicalize, and instruct lone wolf terrorists.
- Countering online terrorism (Von Knop and Weimann 2008, Weimann 2012b): These studies examined various countermeasures and their efficiency, including the first introduction of the concept of "noise" into the model of countering online terror.

These publications reported the project's findings, documenting trends and changes in terrorists' online presence. They highlighted terrorists' fast adaption of new online technologies from Facebook and YouTube to Google Earth, their use of new persuasion and radicalization tactics (e.g., the use of "emarketing" and "narrowcasting" to direct appeals to target online groups like children, women, and "diaspora communities"), and their concern with cyberterrorism (attacking computer networks instead of using them). Many of these findings and publications were presented at congressional hearings and Senate subcommittees, the Department of Homeland Security, the Department of State, and various military and counterterrorism agencies in the United States.

Part I

Terrorism Enters Cyberspace

Chapter 1

Terrorism Enters Cyberspace

Terrorism as Communication

When one says "terrorism" in a democratic society, one also says
"media." For terrorism by its very nature is a psychological weapon
which depends upon communicating a threat to a wider society.
This, in essence, is why terrorism and the media enjoy a symbiotic
relationship.

—Paul Wilkinson, *Terrorism versus Democracy* (2001, 17)

In recent decades, the world has witnessed the emergence and proliferation
of media-wise terrorism. Modern terrorists have become exposed to new
opportunities for exerting mass psychological impacts as a result of technological advances in communications. The emergence of mass-mediated
terrorism has led several communication and terrorism scholars to recon-
ceptualize modern terrorism within the framework of symbolic communication theory (Von Knop and Weimann 2008). As Brian Jenkins (1975, 4)
concluded in his analysis of international terrorism:

Terrorist attacks are often carefully choreographed to attract the attention of the electronic media and the international press. Terrorism is aimed at the people watching, not at the actual victims. Terrorism is a theater.

Indeed, modern terrorism can be understood in terms of the production requirements of theatrical engagements (Weimann and Winn 1994): meticulous attention paid to script preparation, cast selection, sets, props, role-playing, and minute-by-minute stage management. And just like compelling stage plays or ballet performances, the media orientation of terrorist activity requires careful attention to detail in order to be effective. The victim is, after all, only "the skin on a drum beaten to achieve a calculated impact on a wider audience" (Schmid and de Graaf 1982, 14). The growing importance of publicity and mass media to terrorist organizations was revealed both in the diffusion of media-oriented terrorism and in the tactics of modern, media-minded terrorists (Weimann and Winn 1994).

Paralleling the growth in technology-driven opportunities was the effort made by terrorists themselves to hone their communications skills. As one of the terrorists who orchestrated the attack on the Israeli athletes during the 1972 Munich Olympic Games testified:

We recognized that sport is the modern religion of the Western world. We knew that the people in England and America would switch their television sets from any program about the plight of the Palestinians if there was a sporting event on another channel. So we decided to use their Olympics, the most sacred ceremony of this religion, to make the world pay attention to us. We offered up human sacrifices to your gods of sport and television. And they answered our prayers. (Quoted in Dobson and Paine 1977, 15)

The "theater of terror" productions became more frequent and more powerful. Brigitte Nacos (2002, 2003) has noted that since the first World Trade Center bombing in 1993 and the Oklahoma City bombing in 1995, the world has entered into a new age of megaterrorism, and the new age of terrorism is a more powerfully media-oriented production than ever before. The most powerful and violent performance was the al-Qaeda attacks on American targets on September 11, 2001 (9/11). In November 2001, shortly after the 9/11 attacks, al-Qaeda leader Osama bin Laden discussed the twin attacks. Referring to the suicide terrorists, whom he called "vanguards of Islam," bin Laden marveled: "Those young men said in deeds, in New York and Washington, speeches that overshadowed other speeches made everywhere else in the world. The speeches are understood by both Arabs and non-Arabs, even Chinese."[1]

From the "theater of terror" perspective, the 9/11 attacks were a perfectly choreographed production aimed at American and international audiences. Although the theater metaphor remains instructive, it has given way to that of terrorism as a global television spectacular with "live" breaking news; watched by international audiences, terrorism transcends the boundaries of theatrical events. In the past, most if not all acts of terrorism resulted in a great deal of publicity in the form of news reporting. However, the 9/11 attacks introduced a new level of mass-mediated terrorism because of the choices that the planners made with regard to the method, target, timing, and scope of the operation.

The growing use and manipulation of modern communications by terrorist organizations have led governments and several media organizations to consider certain steps in response. These responses included limiting terrorists' access to the conventional mass media, reducing and censoring news coverage of terrorist acts and their perpetrators, and minimizing the terrorists' capacity for manipulating the media. However, new media technologies, computer-mediated communication, and the Internet allow terrorist organizations to transmit messages more easily and freely than through any other means of communication. Terrorist groups and organizations have identified the Internet as an important communication platform. This cheap and global communication tool allows them to enhance their communication strategies.

Online Terrorism

Conflicts on the ground are echoed, as one can imagine, in cyberspace. . . . Cyberspace offers even more fertile territory for sabotage, misinformation, and what in the clichéd formulation is termed the war over mind.

—Vinay Lal, "Terror and Its Networks:
Disappearing Trails in Cyberspace" (2002)

---

[1] The quotes are taken from the translations of a videotape, presumably made in mid-November 2001, in Afghanistan (Michael and Wahba 2001).

Paradoxically, the most innovative network of communication developed by modern Western societies—the Internet, with its numerous online networking platforms—now serves the interests of the greatest foe of the West: international terrorism. Since the late 1980s, the Internet has proven to be a fast and efficient platform for the flow of communication, reaching an ever-growing worldwide audience. The constant development of increasingly sophisticated technologies has created a huge network with a global reach and relatively low barriers to entry. The benefits of online communication are numerous, starting with its unique suitability for sharing information and ideas. Yet the same technologies and platforms that facilitate such communication can also be exploited for the purposes of terrorism and extremism (United Nations Office on Drugs and Crime [UNODC] 2012, 3). This emerging pattern of online terrorism, using and abusing the Internet for terrorist purposes, creates both challenges and opportunities in the endless war against terrorism.

The growing presence of modern terrorism on the Internet is at the nexus of two key trends: the democratization of communications driven by user-generated content on the Internet, and the growing awareness of modern terrorists of the potential of using the Internet as a tool for their purposes. Thus, the Internet has long been a favorite tool of terrorists. Decentralized and providing almost perfect anonymity, it cannot be subjected to control or restriction, and allows access to anyone who wants it. Large or small, terrorist groups have their own websites, and through this medium they spread propaganda, raise funds, and seduce potential newcomers. Additionally, terrorists radicalize audiences, recruit and train members, communicate and conspire with each other, and plan and launch attacks online. Besides thousands of terrorist websites, modern terrorists rely on chatrooms, e-groups, forums, social networking, and online platforms that include You-Tube, Facebook, and Twitter (Von Knop and Weimann 2008; Weimann 2006a, 2011b, 2012b).

In the late 1990s, terrorist movements made their first appearance on the Net (Tsfati and Weimann 2002). In July 2004, the independent National Commission on Terrorist Attacks upon the United States (the 9/11 Commission) released its findings in a 570-page report. In pursuing its mandate, the commission reviewed more than 2.5 million pages of documents and interviewed more than 1,200 individuals in 10 countries. The report starts with a short conclusion: "We learned about an enemy who is sophisticated, patient, disciplined, and lethal. The enemy rallies broad support in the Arab and Muslim world by demanding redress of political grievances, but its hostility toward us and our values is limitless" (National Commission on Terrorist Attacks upon the United States 2004, xvi). It points to the use of modern communication technologies in planning and executing the 9/11 attacks: "Terrorists, in turn, have benefited from this same rapid development of communication technologies. . . . The emergence of the World Wide Web has given terrorists a much easier means of acquiring information and exercising command and control over their operations" (National Commission on Terrorist Attacks upon the United States 2004, 88). It also highlights the ways in which the al-Qaeda operatives used the Internet to design and execute their attacks, including searching the Web for information on US flight schools, using Internet communications (such as equipping the hijackers with email accounts and coordinating the attackers' actions through emails), downloading anti-American webpages, and learning about flights from the Internet. Given this compatible relationship between the Internet and modern terrorism, it is inevitable that terrorists will use and abuse these new electronic communications and capabilities.

The interactive capabilities of the Internet, such as social networking sites, video-sharing sites, and online communities, allow terrorists to assume a proactive position. Instead of waiting for websurfers to come across their websites and propaganda materials, terrorists can now lure targeted individuals to the sites. Online social networking provides terrorists with an ideal platform to attract and seduce, teach and train, radicalize and activate individuals all over the world. The Internet has provided terrorists with a whole new virtual realm to conduct their most sinister transactions. As numerous studies have revealed, most of the recent terrorists involved in the planning of attacks or the attacks themselves were radicalized, recruited, trained, and even launched online, a process that emerged in the early 2000s and has gained momentum ever since (Weimann 2012a). In his 2009 Council on Foreign Relations report on online terrorism, Eben Kaplan (2009) concluded:

> Terrorists increasingly are using the Internet as a means of communication both with each other and the rest of the world. By now, nearly everyone has seen at least some images from propaganda videos published on terrorist sites and rebroadcast on the world's news networks. . . . The Internet is a powerful tool for terrorists, who use online message boards and chat rooms to share information, coordinate attacks, spread propaganda, raise funds, and recruit.

As an illustrative example, one can look at the case of Anwar al-Awlaki, the online preacher of "jihadism," or jihadi terrorism.[2] Al-Awlaki, a radical American Muslim cleric of Yemeni descent, was linked to a series of attacks and plots across the world, from the 9/11 attacks to the on-base shootings at Fort Hood, Texas, in November 2009. He has also been linked to the recruitment of a Nigerian terrorist accused of attempting to blow up Northwest Airlines Flight 253 as it flew into Detroit on December 25, 2009. Al-Awlaki's overt endorsement of violence as a religious duty in his online sermons and YouTube videos is believed to have inspired new recruits to Islamist militancy. As Evan Kohlmann (2009, 2) concluded in his 2009 *Foreign Policy* article on lone wolf terrorism: "Perhaps this is the most frustrating aspect of transnational vendors of hate and mayhem like al Awlaki. The Internet has inadvertently become a powerful tool in their hands, offering easy access to an interactive virtual universe where they can mobilize vulnerable, unstable people around the world and incite them to carry out acts of violence."

Terrorism has changed its face, and so has the Internet and its platforms. In December 2011, the US House of Representatives Subcommittee on Counterterrorism and Intelligence produced the report *Jihadist Use of Social Media—How to Prevent Terrorism and Preserve Innovation.* Subcommittee chairman Patrick Meehan's introductory statement in the report assessed the current situation:

> For years, terrorists have communicated online, sharing al Qaeda propaganda or writing in online forums dedicated entirely to the prospect of Islamist terrorism. But they have recently evolved with technological changes, utilizing social media sites such as Facebook and YouTube to enhance their capabilities. In the same places the average person posts

---

[2] Jihad is a rather complicated Islamic term, open to various interpretations. In Arabic, the word translates as a noun meaning "struggle." Within the context of the Quran and classical Islam, particularly in Shiite beliefs, it refers to the important religious duty to struggle against those who do not believe in Allah. However, there are two commonly accepted meanings of jihad: an inner spiritual struggle to fulfill one's religious duties ("greater jihad") and an outer physical struggle against the enemies of Islam ("lesser jihad"). The physical struggle can be violent or nonviolent, and proponents of the violent form translate jihad as "holy war," calling for military assault on nonbelievers. Middle East historian Bernard Lewis argues that "the overwhelming majority of classical theologians, jurists, and traditionalists (specialists in the hadith) understood the obligation of jihad in a military sense," and maintains that for most of the recorded history of Islam the word jihad was used in a primarily military sense (Lewis 1988, 72). "Jihadism" in this sense covers both guerrilla warfare and the international Islamic terrorism substantially represented by the al-Qaeda network. The terms "jihadists" and "mujahideen" are associated with those who follow the more radical and violent interpretation of global jihad.

photos and communicates with friends and family, our enemies distribute videos praising Osama bin Laden. . . . These examples highlight the incredible challenge posed by terrorists engaging online. The Internet was designed to ease communication, and it must stay that way. However, we cannot ignore the reality that we have been unable to effectively prevent jihadi videos and messages from being spread on popular social media websites like YouTube and Facebook.[3]

The so-called information revolution, with the dramatic rise of the Internet and online platforms, has clearly been of growing societal, political, and cultural significance. For terrorists and extremists, these new online platforms offer unique opportunities and capabilities to communicate, collaborate, and convince.

The Advantages of Cyberspace for Modern Terrorism

> The greatest advantage [of the Internet] is stealth. [Terrorists] swim in an ocean of bits and bytes.
>
> —John Arquilla, professor of defense analysis,
> Naval Postgraduate School (quoted in Kaplan 2009)

By its very nature, the Internet is in many ways an ideal arena for terrorist and other extremist group activities. The great virtues of the Internet—its ease of access, lack of regulation, vast potential audiences, fast flow of information, and so forth—have been converted into the advantage of groups committed to terrorizing societies to achieve their goals.

The anonymity offered by the Internet is very attractive for modern terrorists. Because of their extremist beliefs and values, terrorists require anonymity to exist and operate in social environments that may not agree with their particular ideological views or their activities. The Internet provides this anonymity, in addition to easy, universal access with the options of posting messages, sending emails, or uploading or downloading information. Once these goals have been accomplished, the Internet allows terrorists to disappear into the dark. As a result, terrorists have become increasingly sophisticated at exploiting online platforms for anonymous communications. For instance, terrorists may use a simple online email account for electronic

---

[3] Statement of Rep. Patrick Meehan, chairman of the House Subcommittee on Counterterrorism and Intelligence, December 6, 2011, http://homeland.house.gov/sites/homeland.house.gov/files/12 -06-11%20Meehan%20Open.pdf.

"dead dropping" of communications. This refers to the creation of a draft message, which remains unsent and therefore leaves minimal electronic traces, but which may be accessed from any Internet terminal worldwide by multiple individuals who have the relevant password. This practice permits messages to reach a broad but anonymous audience, and to be edited and deleted with little evidence that they ever existed.

Emerging technologies have made it easier and cheaper for terrorists to communicate online and increasingly difficult for authorities to monitor these communications. Email, chat rooms, mobile phones, SMS (Short Message Service) text communications, VoIP (Voice-over-Internet Protocol) communications, social media, online video sharing, virtual worlds, and micro-blogging sites not only are ideal platforms for terrorists, but also create enormous volumes of communication flow, thus allowing terrorists to hide among the noise. In addition to the numerous platforms, alternative communication devices have proliferated: conversations, videos, or images can be placed on mobile phones, laptop computers and tablets, social media, websites, consumer electronics such as portable music players, or gaming devices, many of which can be obtained without providing any subscriber information. An abundance of more sophisticated measures and technologies also increase the difficulty of identifying the originator, recipient, or content of terrorist online communications. These include encryption tools and anonymizing software that are readily available online for download.

These advantages have not gone unnoticed by terrorist organizations, regardless of their specific orientation. Islamists and Marxists, nationalists and separatists, fundamentalists and extremists, racists, and anarchists all find the Internet alluring. Today, all active terrorist organizations maintain websites, and many maintain more than one website and use several different languages. As the following illustrative list shows, these organizations and groups come from all corners of the globe:

- *Africa:* Al-Qaeda in the Islamic Maghreb (AQIM); the Muslim Brotherhood in Egypt; al-Shabaab in Somalia, and numerous al-Qaeda affiliates including groups in Algeria, Libya, and Sudan.
- *Asia:* Aum Shinrikyo (Supreme Truth) in Japan, the Japanese Red Army, Hizb-ul-Mujahideen in Kashmir, the Liberation Tigers of Tamil Eelam in Sri Lanka, the Islamic Movement of Uzbekistan, the Moro Islamic Liberation Front in the Philippines, the Pakistan-based Lashkar-e-Taiba, the Chechnya rebel movement, and al-Qaeda affiliates in Asian countries.

- *Europe:* Armata Corsa (the Corsican Army), the Basque ETA move-ment, the Real Irish Republican Army, and several al-Qaeda affiliates in European countries.
- *Latin America:* Túpac Amaru and Sendero Luminoso (Shining Path) in Peru, the Colombian National Liberation Army, the Armed Revo-lutionary Forces of Colombia, and al-Qaeda affiliates in South American countries.
- *Middle East:* Hamas (the Islamic Resistance Movement); the Lebanese Hezbollah (Party of God); the al-Aqsa Martyrs' Brigades; the Palestinian Fatah Tanzim; the Popular Front for the Liberation of Palestine; the Palestinian Islamic Jihad; the Kahane Lives movement in Israel; the Mojahedin-e Khalq (People's Mojahedin of Iran); Ansar al-Islam (Supporters of Islam) in Iraq; the Kurdish Workers' Party; the Turkish-based Popular Democratic Liberation Party–Front; and numerous al-Qaeda affiliates, including groups in Gaza, Lebanon, Saudi Arabia, the Sinai Peninsula, Syria, and Yemen.

Terrorists' Uses of Cyberspace

Terrorism and the Internet are related in several ways. The Internet has become a forum for terrorist groups and individual terrorists to spread their messages of hate and violence and to communicate with one another and with their sympathizers. Moreover, individuals and groups may attempt to attack computer networks—with the intent, for instance, of bringing down airplanes, ruining critical infrastructure, destabilizing the stock market, or revealing state secrets—in an act that has become known as cyberterrorism or cyberwarfare. At this point, terrorists are using cyberspace mostly for propaganda and communication purposes rather than attacking purposes, but as this book will discuss, cyberterrorism is certainly on the terrorists' agenda and is likely to become their new mode of operation.

Several studies have monitored terrorists' varied uses of the Internet (e.g., UNODC 2012; Weimann 2005a, 2006a). These studies identified numerous (and sometimes overlapping) categories of use, which can be grouped into two kinds: communicative and instrumental. Communicative uses include spreading propaganda, launching psychological warfare campaigns, secur-ing internal communications, and radicalizing recruits (Coll and Glasser 2005; Conway 2002, 2005; Cronin 2006; Glasser and Coll 2005; Labi 2006; Lynch 2006; Rogan 2006; Swartz 2005; Talbot 2005; Thomas 2003; Vatis 2001; Weimann 2006a, 2006b). Instrumental uses include online teaching and training of terrorists, and establishing "virtual training camps" for future assailants (Weimann 2005a, 2006b, 2007a, 2007b, 2008b, 2009d). The most popular terrorist-related uses of online platforms are (1) psychological warfare, (2) propaganda, (3) online indoctrination, (4) recruitment and mobilization, (5) data mining, (6) virtual training, (7) cyberplanning and coordination, and (8) fund-raising.

PSYCHOLOGICAL WARFARE

Terrorism has often been conceptualized as a form of psychological warfare, and terrorists have certainly sought to wage such a campaign through the Internet. They may do so in several ways. For instance, they can use the Internet to spread threats intended to distill fear and helplessness, and to disseminate horrific images of recent actions, such as the brutal 2002 murder of the kidnapped American journalist Daniel Pearl by his Pakistani captors, a recording of which was replayed on several terrorist websites. Other means include scary warnings about incoming attacks and threats targeting specific nations, cities, or populations. The Internet, an uncensored medium that carries stories, pictures, threats, or messages regardless of their validity or potential impact, is particularly well suited to allowing even a small group to amplify its message and exaggerate its importance and the threat it poses. Very often, this is done by feeding the "conventional" media rather than through direct exposure: journalists often cite the threats and declarations taken from online terrorist publications, thus magnifying their impact (Weimann 2008f, 65).

Al-Qaeda has consistently claimed on its websites that the 9/11 destruction of the World Trade Center has inflicted both psychological and concrete damage on the US economy. The attacks on the Twin Towers are depicted as an assault on the trademark of the US economy, and evidence of their effectiveness is seen in the weakening of the dollar, the decline of the US stock market after 9/11, and the supposed loss of confidence in the US economy both within the United States and abroad. Parallels are drawn with the decline and ultimate demise of the Soviet Union. One of Osama bin Laden's publications, posted online, declared that "America is in retreat by the Grace of Almighty and economic attrition is continuing up to today. But it needs further blows. The young men need to seek out the nodes of the American economy and strike the enemy's nodes" (quoted in Hoffman 2006a, 124).

PROPAGANDA

One of the primary uses of online communication by terrorists is for the dissemination of propaganda (Minei and Matusitz 2011, 2012; Weimann 2006a, 2012b). This generally takes the form of multimedia communications providing ideological, political, or religious explanations, justifications, or promotion of terrorist activities. These communications may include online messages, streaming videos, social media messages, and even video games developed by terrorist organizations. The Internet has significantly expanded the opportunities for terrorists to garner publicity. Until the advent of the Internet, terrorists' hopes of winning publicity for their causes and activities depended on attracting the attention of television, radio, or print media networks. These traditional media have "selection thresholds"—multistage processes of editorial selection—that terrorists often cannot reach. No such thresholds, of course, exist on the terrorists' own websites. Many terrorists now have direct control over the content of their message, which offers further opportunities to shape how they are perceived by different target audiences and allows them to manipulate their own image in addition to that of their enemies (Zanini and Edwards 2001, 42). According to a 2012 UNODC report, "The fundamental threat posed by terrorist propaganda relates to the manner in which it is used and the intent with which it is disseminated. Terrorist propaganda distributed via the Internet covers a range of objectives and audiences. It may be tailored, inter alia, to targeted audiences that range from potential or actual supporters, to opponents of an organization or shared extremist belief, to direct or indirect victims of acts of terrorism, or to the international community or a subset thereof" (UNODC 2012, 4).

Most terrorist online propaganda does not celebrate their violent activities. Instead, regardless of the terrorists' agendas, motives, and location, terrorist messages emphasize two issues: the restrictions placed on freedom of expression, and the plight of comrades who are now political prisoners (Weimann 2005c). These issues resonate powerfully with their own supporters and are also calculated to elicit sympathy from Western audiences that cherish their freedoms of expression and frown on measures to silence political opposition. Enemy publics, too, may be targets for these complaints insofar as the terrorists, by emphasizing the antidemocratic nature of the steps taken against them, try to create feelings of unease and shame among their foes. The terrorists' protest at being muzzled is particularly well suited to the Internet, which for many users is the symbol of free, unfettered, and uncensored communication.

ONLINE INDOCTRINATION

Modern terrorists have made the Internet an instrument for radicalization and indoctrination. Many of the recent terrorist attacks in Europe, North Africa, and the Middle East were executed by people who had been indoctrinated through the Internet. Recruitment, radicalization, and incitement to terrorism should be viewed as points along a continuum (UNDOC 2012, 6). Radicalization refers primarily to the process of indoctrination that often accompanies the transformation of recruits into individuals determined to act with violence based on extremist ideologies. Terrorist groups can use a variety of online platforms to indoctrinate potential recruits, ranging from personal emails and online chats to seductive videos and social media.

Along with indoctrination, there is also a process known as "online self-radicalization." This is a newer type of terrorist, the homegrown extremist who cultivates his or her views online. The Boston Marathon bombing on April 15, 2013, involved two brothers, Dzhokhar and Tamerlan Tsarnaev, who used two pressure-cooker bombs to kill three people and injure an estimated 264 others. According to Federal Bureau of Investigation interrogators, the Tsarnaev brothers were motivated by extremist Islamic beliefs through their exposure to online messages of radical Islam (Weimann 2014b). In recent years, the brothers became interested in Islamic extremism. A video on their YouTube channel, for example, featured the radical cleric Feiz Mohammad, who lives in Australia. Tamerlan Tsarnaev downloaded a significant amount of jihadist material from the Internet, including a book about "disbelievers" with a foreword by the radical cleric Anwar al-Awlaki. In addition, he downloaded the first volume of the al-Qaeda-produced online magazine *Inspire*, which offered detailed instructions for building bombs with pressure cookers, shrapnel, and explosive powder from fireworks. Dzhokhar appears to have been influenced by his older brother Tamerlan, but he, too, apparently was radicalized online and reportedly confessed in a note that the bombings were meant as retribution for the American-led wars in Iraq and Afghanistan. Radical Islamic websites often use US military operations in Islamic countries as justification for terrorist violence (Siddiqi and Kaleem 2013).

RECRUITMENT AND MOBILIZATION

The Internet and advanced technology provide powerful tools for recruiting and mobilizing group members through integrated communications. In addition to seeking converts by using the full panoply of website technologies (e.g., audio, digital video) to enhance the presentation of their message, terrorist organizations capture information about the users who browse their websites. Users who seem most interested in the organization's cause or appear well suited to carrying out its work are then contacted. Recruiters may also use more interactive Internet technology to roam online chat rooms, Facebook, Twitter, and other platforms, looking for receptive members of the public, particularly young people. The reach of these online outlets provides terrorist organizations and sympathizers with a global pool of potential recruits. The virtual forums offer an open venue for recruits to learn about and provide support to terrorist organizations, and they promote engagement in direct actions (Denning 2010; Gerwehr and Daly 2006; Weimann 2007a).

Global counterterrorism efforts drove many terrorist groups, including al-Qaeda, to call on their devotees in the West to carry out smaller-scale solo attacks, mainly by providing them with online education to teach them how to do so. "I strongly recommend all of the brothers and sisters coming from the West to consider attacking America in its own backyard," wrote Samir Khan, an American who joined al-Qaeda's Yemen branch and emerged as a fervent advocate of homegrown, do-it-yourself terrorism before he was killed in an American drone strike in September 2011 (Shane 2013). In recent years, al-Qaeda propagandists have "made a particular effort to recruit lonely people who are looking for a cause," said Jerrold Post, a former US Central Intelligence Agency (CIA) psychiatrist now at George Washington University and the author of *The Mind of the Terrorist: The Psychology of Terrorism from the IRA to al-Qaeda* (quoted in Cobb 2013). Post points to, among others, Major Nidal Hasan, the Army psychiatrist accused of killing 13 people in the 2009 Fort Hood shooting. Hasan was held up as an example for others to follow in a two-part video released by al-Qaeda's core group in Pakistan in June 2011. This video, titled "You Are Only Responsible for Yourself," urged Muslims in the West to stage attacks without waiting for orders from abroad.

Terrorist recruitment-oriented communication is often tailored to appeal to vulnerable and marginalized groups in society. Often, this propaganda of recruitment and radicalization capitalizes on an individual's sentiments of injustice, alienation, or humiliation (European Commission 2008; Weimann 2008b). As discussed in chapter 2, terrorists have become more sophisticated with their usage of "narrowcasting," or propaganda that targets specific subpopulations according to demographic factors (such as age or gender) as well as social or economic circumstances (Weimann 2008b, 2008f, 2009b).

Cyberspace may be an ideal platform for recruiting children and youth, who comprise a high proportion of users. Propaganda disseminated via the Internet with the aim of recruiting minors may take the form of cartoons, popular music videos, or computer games. These contents often mix cartoons and children's stories with messages promoting and glorifying acts of terrorism, such as suicide attacks. Similarly, some terrorist organizations have designed online video games to be used as recruitment and training tools. Such games may promote the use of violence, rewarding virtual successes (Weimann 2008b). For example, Al-Qaeda in the Islamic Maghreb is changing its strategy to target children at an early age to lure them to its radical ideology. To do this, beginning in March 2013 the group started to use new methods deemed to be more capable of attracting children's attention, such as video games that include a clear strategy to show the group's ability to win wars against international forces. On its website, AQIM published a computer game called "Muslim Mali," in which players operate a military aircraft carrying AQIM's black flag to attack and destroy French aircraft in the Sahara, where battles are raging against the terrorists in northern Mali. The website says that the game displays the message "Congratulations, you have become martyrs!" in lieu of "Game Over" when a player loses all their lives.

DATA MINING

The Internet is a vast source of information on all topics; in fact, it may be viewed as a vast digital library. The World Wide Web alone offers billions of pages of information, much of it free—and much of it of interest to terrorist organizations. Terrorists can use the Internet to gather information that may be relevant to their cause or to future operations. They can learn a wide variety of details about prospective targets, such as transportation facilities, nuclear power plants, public buildings, airports, and ports. They can acquire satellite images, maps, and blueprints of these targets, and they can even learn about counterterrorism measures. In his book *Black Ice: The Invisible Threat of Cyber-Terrorism*, Dan Verton (2003, 184) argued that "al-Qaeda cells now operate with the assistance of large databases containing details of potential targets in the U.S. They use the Internet to collect intelligence on those targets, especially critical economic nodes, and modern software enables them to study structural weaknesses in facilities as well as predict the cascading failure effect of attacking certain systems." According to former Secretary of Defense Donald Rumsfeld, speaking on

January 15, 2003, an al-Qaeda training manual recovered in Afghanistan tells its readers that "[u]sing public sources openly and without resorting to illegal means, it is possible to gather at least 80 percent of all information required about the enemy" (quoted in Thomas 2003).

In addition to information provided by and about the armed forces, the free availability of information on the Internet about the location and operation of nuclear reactors and related facilities was of particular concern to the authorities after 9/11. Roy Zimmerman, director of the Nuclear Regulatory Commission's (NRC) Office of Nuclear Security and Incident Response, said that the 9/11 attacks highlighted the need to safeguard sensitive information. In the days immediately after the attacks, the NRC took its website offline, and when it was restored weeks later it had been purged of more than a thousand sensitive documents (Ahlers 2004). The measures taken by the NRC were not exceptional. According to a report produced by OMB Watch, since 9/11, thousands of documents and tremendous amounts of data have been removed from US government sites.[4] When US forces in Afghanistan discovered al-Qaeda computers with US dam drawings stored on them, the Army Corps of Engineers stopped posting engineering project designs as a part of contract solicitations.

Recently, concern has arisen over the availability of satellite images from Google Earth on the Internet. The images, which Google Earth updates about every 18 months, are a patchwork of aerial and satellite photographs, and their relative sharpness varies. For a brief period, photos of the White House and adjacent buildings that the United States Geological Survey provided to Google Earth showed up with certain details obscured, because the government had decided that showing places like rooftop helicopter landing pads was a security risk (Hafner and Rai 2005). Terrorists do use Google Earth services: detailed Google Earth images of British military bases were found in the homes of Iraqi insurgents; Hamas terrorists from Gaza have used Google Earth to aim their rockets when targeting Israeli towns; and the terrorists that attacked various locations in south Mumbai, India, in November 2008 used digital maps from Google Earth to learn their way around (Harding 2007, Schneier 2009). Investigations by the Mumbai police, including the interrogation of one captured terrorist, suggest that the terrorists were highly trained and that they used technologies such as

---

[4] The Center for Effective Government was founded as OMB Watch in 1983, with a primary focus on making the work of executive branch agencies more transparent and open to citizen input. The report is quoted in Declan McCullagh (2003).

satellite phones and global positioning systems (GPS) linked to Google Earth satellite images. In the planned attempt by terrorists to blow up fuel tanks at New York's John F. Kennedy International Airport in 2007, court records indicate that the plotters utilized Google Earth to obtain detailed aerial photographs of their intended target (Buckley and Rashbaum 2007). The program is seen as superior to maps because it is more up-to-date and gives precise locations for potential targets.

Finally, incidents such as the 2011 "Wikileaks" disclosures, which leaked more than 250,000 diplomatic cables, provide government assessments on the state of terrorist organizations, their plans and intentions, and therefore reveal the extent of their knowledge. In August 2013, a leak concerning an al-Qaeda plot to attack US embassies in the Middle East apparently undermined US intelligence by prompting terrorists to change their methods of communicating. Citing US officials and experts, the *New York Times* reported on September 29, 2013, that the details of the leaked al-Qaeda plot in August "caused more immediate damage to American counterterrorism efforts" than documents disclosed by former National Security Agency contractor Edward Snowden (Schmitt and Schmidt 2013).

VIRTUAL TRAINING

The Internet has become a valuable tool for terrorist organizations, not just for indoctrination, propaganda, and recruitment, but also as a home for virtual training camps for the practical application of terrorism. Thousands of new pages of terrorist manuals, instructions, and rhetoric are published on the Internet every month. Some experts have referred to the Internet as a "terrorist university," a place where terrorists can learn new techniques and skills to make them more effective in their attack methodologies (United Nations Counter-Terrorism Implementation Task Force 2011, 20). Continuous attacks on al-Qaeda camps in Afghanistan and elsewhere have forced the terror group to move its base of operations and training camps into cyberspace. "It is not necessary . . . for you to join a military training camp, or travel to another country . . . you can learn alone, or with other brothers, in [our arms] preparation program," al-Qaeda leader Abu Hajir al-Muqrin announced in 2004. Today, as attacks on al-Qaeda and other terrorist groups intensify, their reliance on the Internet for launching instruction and training campaigns appears to have expanded significantly and become more sophisticated. Moreover, the online "courses" have become more developed and technologically advanced.

Online magazines (such as al-Qaeda's *Inspire*) and many terrorist web-sites also provide training and ideas for terrorist attacks, and the development and widespread availability of the Internet has made it possible to create readily accessible "virtual training camps" (Amble 2012, Weimann 2009a). Readily available online documents include the "Mujahideen Poisons Handbook," an instructive manual which contains various "recipes" for homemade poisons and poisonous gases. Similar information on hostage taking, bomb making, and guerrilla tactics is also available in a wide variety of other sources such as the "Anarchist Cookbook" and the "Sabotage Handbook." The 600-page "Encyclopedia of Jihad" is also widely available online and includes chapters such as "how to kill," "explosive devices," "manufacturing detonators," and "assassination with mines."

The convenience of access and anonymity that cyberspace provides has done away with the necessity for training in formal training camps for specific tactics being used by the terrorists in many conflict zones. Al-Qaeda's first online instruction was in the form of a colorful online magazine called *Al Battar Training Camp. Al Battar* takes its name from the "Sword of the Prophets," a weapon once owned by the Prophet Muhammad. In early 2004, al-Qaeda published online the first issue of *Al Battar.* The introduction to the issue states:

Because many of Islam's young people do not yet know how to bear arms, not to mention use them, and because the agents of the Cross are hobbling the Muslims and preventing them from planning [jihad] for the sake of Allah—your brothers the Mujahideen in the Arabian peninsula have decided to publish this booklet to serve the Mujahid brother in his place of isolation, and he will do the exercises and act according to the military knowledge included within it. . . . The basic idea is to spread military culture among the youth with the aim of filling the vacuum that the enemies of the religion have been seeking to expand for a long time.

Later, the issue promoted online training: "O Mujahid brother, in order to join the great training camps you don't have to travel to other lands. Alone, in your home or with a group of your brothers, you too can begin to execute the training program." In November 2008, the SITE Intelligence Group reported that al-Nusra Media Battalion, a jihadist media group, had compiled into a single file a collection of explosives manuals totaling over a thousand pages and posted the file on jihadist forums. This collection, entitled "The Encyclopedia of Weapons and Explosives, First Part," includes manuals that are frequently distributed on jihadist forums, such as those written by explosives expert Abdullah Dhu al-Bajadin and distributed by the Islamic Media Center. These manuals provide instructions for a range of compounds and equipment within this field, including mobile phones for remote detonation, chemical explosives, detonators, and placement of these materials to strike a specific target (SITE Monitoring Service 2008).

The use of new technologies for launching terrorist attacks is a frequent topic discussed in these forums and chat rooms. Thus, videos containing instructions to prepare explosives, optimized for viewing on mobile phones, were posted on the al-Fallujah jihadist forum on October 27, 2008. The videos comprise the explosives series, "Lessons in How to Destroy the Cross," which received much attention by jihadists when it was posted on al-Ekhlaas, an al-Qaeda-affiliated forum. The mobile phone format is a standard by which al-Qaeda–affiliated groups such as the Islamic State of Iraq, AQIM, and the As-Sahab Foundation for Islamic Media Publication[5] distribute their videos. They comprise approximately 19 hours of footage. Fifteen of the lessons focus on particular materials, such as TNT, picric acid, nitroglycol, sodium acid, and ammonium nitrate. The other 10 videos involve seminars held on sensitive, semisensitive, and nonsensitive substances and detonators, showing an individual leading the course using a whiteboard, illustrations, and various objects. Jihadist forums host discussions on the use of modern online platforms such as Twitter, Facebook, YouTube, and others.

Finally, the most popular online training manual is the glossy, ultramodern *Inspire* magazine. *Inspire* is an English-language online magazine published by Al-Qaeda in the Arabian Peninsula (AQAP). Numerous international and domestic Islamist extremists have been influenced by the magazine and, in some cases, reportedly used its bomb-making instructions to carry out attacks. The magazine is an important branding tool for all al-Qaeda branches, franchises, and affiliates (Merriam 2011). The magazine promotes "open source jihad," a necessary tactic as al-Qaeda leadership has steadily vanished in the 10 years since 9/11. With its leaders either dead or in jail, al-Qaeda had to consider new ways to attack its enemies. This caused a shift away from al-Qaeda's traditional terrorist attacks and toward simple attacks by individuals using common items for weapons. The summer 2010 issue advised making a pressure-cooker bomb using everyday materials ("How to make a bomb in the kitchen of your mom"), a method

---

[5] "As-Sahab" means "the cloud" in Arabic.

used by the 2013 Boston Marathon bombers. The eleventh issue of *Inspire*, published online in June 2013, devoted almost all of its forty-odd pages to glorifying what it calls the "BBB": the "Blessed Boston bombings." One article, "Inspired by *Inspire*," is illustrated by a flaming iPad with a copy of the magazine on its screen. The main takeaway from the June 2013 issue is that its editors are unabashedly pleased that copies of their magazine were found in the Tsarnaev brothers' house.

CYBERPLANNING AND COORDINATION

Not only do terrorists use the Internet to learn how to build bombs, but they also use it to plan and coordinate specific attacks. Al-Qaeda operatives relied heavily on the Internet for the planning and coordination of the 9/11 attacks. Numerous messages that had been posted in a password-protected area of a website were found by federal officials on the computer of arrested al-Qaeda terrorist Abu Zubaydah, who masterminded the 9/11 attacks. To preserve their anonymity, the al-Qaeda terrorists used the Internet in public places and sent messages via public email. Some of the 9/11 hijackers communicated using free web-based email accounts (Weimann 2006d). Many more sophisticated methods exist to ensure anonymity online, including usage of the increasing number of cost-free anonymization services such as the I2P Network and the Tor Project, each of which uses a variety of peer-to-peer and encryption technologies to hide Internet protocol (IP) addresses. These anonymization services utilize a proxy server computer that acts as an intermediary and a privacy shield between the client computer and the rest of the Internet. In effect, the proxy acts on the original user's behalf to protect any personal information from being shared with destination points on the Internet beyond the proxy, and as such, users are able to "spoof" or alter their IP address. These proxy services have increased in sophistication in recent years and now often utilize a peer-to-peer networking approach in order to prevent the user's identity from remaining in any single central third-party site that could disclose the identity.

Hamas activists in the Middle East, for example, use chatrooms to plan operations, and operatives exchange email to coordinate actions across Gaza, the West Bank, Lebanon, and Israel. Instructions in the form of maps, photographs, directions, and technical details of how to use explosives are often disguised by means of steganography, which involves hiding messages inside graphic files. Sometimes, however, instructions are delivered concealed in only the simplest of codes. Mohammed Atta's final message to the other 18 terrorists who carried out the 9/11 attacks is reported to have read: "The semester begins in three more weeks. We've obtained 19 confirmations for studies in the faculty of law, the faculty of urban planning, the faculty of fine arts, and the faculty of engineering" (quoted in Fouda and Fielding 2003, 140). The reference to the various faculties was apparently the code for the buildings targeted in the attacks.

Savvy terrorists have adapted new and advanced communication technologies for their networking, including two emerging technologies: mobile computing technologies and "cloud computing." Most mobile phones now provide easy access to the Internet, and the wide availability of nonregistered SIM cards in many countries allows users to make phone calls, send text messages, and surf the Internet without needing to provide any form of identification. In addition, the wide availability of cloud computing resources, which store information on a shared, accessible online network, means that terrorists are able to host and store their propaganda material, manuals, and all digital content online with little fear of identification or reprisal. In the 2008 terrorist attack on numerous locations in Mumbai, the attackers used all of the most advanced communication technologies. These included handheld GPS devices to plan and perpetrate their attack as well as Google Earth satellite imagery. They received live updates from their handlers on their mobile phones with regard to the location of hostages, especially foreigners (United Nations Counter-Terrorism Implementation Task Force 2011, 32).

FUND-RAISING

The Internet is probably the simplest, easiest to operate, and least costly method of soliciting donations and contributions. Terrorists use a variety of techniques to raise funds online for their activities. Following a popular business trend, many have turned to e-commerce, selling CDs, DVDs, T-shirts, and books as a means of raising cash. An even easier approach is merely to "accept donations," and many terrorist organizations have added links to their sites which advise visitors on how to donate funds electronically via bank transfer. Through the Internet, a terrorist group or its supporters can ask for donations to support its actions directly or can, as is often done, disguise its solicitations as support for charity. Many terrorist organizations also create "charitable organizations" through which they solicit funds, promising to use the money to feed and clothe the poor, though their true intent is to fund acts of violence (Vaccani 2010). Several terrorist organizations have used social networking applications as the latest method for raising money for their activities. Often, the donations are presented as a religious duty and a substitute for joining jihad as an actual warrior.

Almost all groups have used the Internet to solicit donations. Islamist terrorist groups, including Hamas, Lashkar-e-Taiba, and Hezbollah, have also made extensive use of the Internet to raise and transfer needed funds to support their activities (Jacobson 2010b). Hezbollah used to make direct appeals for donations for the "sustenance of the Intifada" on the website of its English-language satellite Al-Manar. The website provided details of the bank account to which donors could send money. Most of the websites linked to Hezbollah, the Palestinian Islamic Jihad, and Hamas openly seek donations to support the families of suicide bombers. The website of the Global Jihad Fund asked for donations to be sent to bank accounts in Pakistan "to facilitate the growth of various jihad movements around the world by supplying them with sufficient funds to purchase weapons and train their individuals." In all these cases, in addition to providing humanitarian aid, such organizations pursue a covert agenda of providing financial and material support to militant groups. Terrorist leaders like Osama bin Laden, Ayman al-Zawahiri, and others have made a number of appeals for money though online speeches, statements, and posted videos. Jihadi leaders often argue that contributing money is similar to physically engaging in jihad. The English-language website of the Pakistan-based group Harakat ul-Mujahideen stated, "Allah gives you the opportunity to take part in the struggle for Muslim rights—jihad. Even if you cannot take part physically in the jihad, you can help us by the means of financial aid."

Another way that terrorist groups are using the Internet to raise funds is through criminal activity. As a 2011 United Nations counterterrorism report stated, "There is substantial evidence that terrorist organizations are using the proceeds from traditional cybercrime, such as online credit card fraud, identity theft and telecommunications fraud to fund their operations" (United Nations Counter-Terrorism Implementation Task Force 2011, 34). Terrorists also use phishing scams to seduce innocent individuals into providing their credit card details. There are numerous examples of this growing trend. The cell that executed the 2004 Madrid train bombing plot, which killed almost two hundred people, partially financed the attack by selling hashish. The terrorists who carried out the July 7, 2005 attacks on the London transportation system were also self-financed, in part through credit card fraud. The al-Qaeda–affiliated Jemaah Islamiyah financed the 2002 Bali bombings, in part, through jewelry store robberies. Key terrorist leaders have specifically encouraged their followers to pursue this path.

For example, Imam Samudra, a former Jemaah Islamiyah operative convicted for his role in the 2002 Bali bombings, wrote a book that included a chapter entitled "Hacking, Why Not?" In that chapter, Samudra urged other jihadists to use credit card fraud and money laundering, and even pointed his readers to specific websites that would help individuals get started and to chat rooms where they could find hacking "mentors" (Jacobson 2010a).

Online (Virtual) Jihad

The Internet has become a major communication channel in the Arab world. In fact, it is the second most important source of information, news, and opinions (see Bunt 2003, 2009). Between 2000 and 2012, Internet connectivity grew by 566 percent worldwide, but figures for the Middle East show a growth of nearly 2,640 percent, while Internet penetration in Pakistan expanded by a staggering 15,000 percent (Miniwatts Marketing Group 2013). Among the earliest jihadist attempts to use the Internet was the establishment of Azzam.com in 1997 by a student at Imperial College, London. Today, most of the terrorist groups on the Internet belong to radical Islamist and jihadi organizations. This should come as no surprise, in light of several revealing studies and especially Gary Bunt's research in his books *Virtually Islamic* (2000), *Islam in the Digital Age* (2003), and *iMuslims: Rewiring the House of Islam* (2009).

Bunt provides a detailed description of the diverse manifestations of the jihadi presence online. He suggests that Islamic organizations have adapted to the digital age by significantly redirecting their resources to the Internet, preferring online communication over traditional channels. This trend is reflected in the volume of militant Islamic materials online and in the growing sophistication of Islamic websites. For example, the presentation of video clips and audio broadcasts on Islamic sites applies some of the most recent developments in computer technology. Bunt also argues that "the Islamic Internet landscape changes frequently, with new sites emerging on a daily basis. Some very proactive players change their content and format regularly, attempting to draw readers to their message(s) in order to establish links or a sense of community" (Bunt 2000, 10). Chatrooms are often unregulated and unmonitored by scholars and clerics, can provide a virtual hangout for teenage and young adult Muslims, and are sometimes rife with anti-*kuffar* (nonbeliever) sentiment. Bunt concludes, "The Internet is clearly important in disseminating a broad range of Islamic political-religious opinions and concerns to a global audience. Thus, many extremist Islamist activists and terrorists now see the Internet as a vital tool" (Bunt 2000, 14). According to Bunt's *iMuslims* (2009), the Internet has profoundly shaped how Muslims perceive Islam and how Islamic societies and networks are evolving and shifting within the twenty-first century. Although these electronic interfaces appear innovative in terms of how the media is applied, much of their content has a basis in classical Islamic concepts, with a historical resonance that can be traced back to the time of the Prophet Muhammad.

One of the most complete reports of the so-called electronic jihad, detailing all the online activities that al-Qaeda supporters are encouraged to conduct, is the book *39 Ways to Serve and Participate in Jihad* by Mohammad bin Ahmad al-Salem, which first appeared online in 2003 (quoted in Kassimeris and Buckley 2010, 448). Among the different activities that can be developed in and through the Internet, propaganda (in its various forms) is the most accessible and doable by grassroots activists without any particular technological skill. For example, the author of the "39 ways" suggests that militants register in forums, post messages strategically in order to keep jihadists topics always at the top of the forum, and support jihadist views.

According to the 2013 New America Foundation report *The State of Global Jihad Online*, the specter of jihadi online radicalization came dramatically to the fore in 2010, largely because of jihadi plots linked to American-born cleric Anwar al-Awlaki (Zelin 2013). But the importance of al-Qaeda communications long precedes al-Awlaki's involvement. As al-Qaeda leader Osama bin Laden remarked in 2002, "It is obvious that the media war in this century is one of the strongest methods; in fact, its ratio may reach 90% of the total preparation for the battles."[6] Much of that jihadi media war now occurs online. In the New America Foundation report, Aaron Zelin (2013, 4–5) highlights four different phases in which jihadi media have been disseminated since 1984:

- Phase 1 (Beginning in 1984): *Khutbas* (sermons), essays/pamphlets, printed magazines, newsletters, and videotaped lectures and/or battle scenes.
  Examples: Abdullah Azzam's tours of mosques in Europe and the United States; a variety of old VHS tapes that came out of Afghanistan, Bosnia, and Chechnya; and jihadi magazines.
- Phase 2 (Beginning in the mid-1990s): Top-down websites: Completely centralized endeavors in which an individual owning a Web domain

---

[6] Quote from "Letter to Mullah Mohammed Omar from Usama bin Ladin," dated June 5, 2002. Located in the United States Military Academy's Combatting Terrorism Center online Harmony Database, Document # AFGP-2002-600321. http://www.ctc.usma.edu/wp-content /uploads/2010/08/AFGP-2002-600321-Trans.pdf.

(often connected directly with jihadi organizations) held complete monopoly over what content was important and would be distributed. Examples: alneda.com and Azzam Publications.

- Phase 3 (Beginning in the mid-2000s): Interactive forums: Administrators of the forums help facilitate and disseminate content on behalf of jihadi organizations, but they are not necessarily directly linked. They post important news items and have the power to delete threads and ban users, allowing them to help steer the online community in a certain direction by preventing users from being exposed to particular content or dissent. At the same time, users can play a role in posting a variety of materials, including their own views on events, and have the ability to converse with like-minded individuals across a wide geographic area.

  Examples: al-Hesbah, al-Ikhlas, al-Fallujah, Ansar, and Shamukh.

- Phase 4 (Beginning in the late 2000s): Social media platforms: A particular individual is in control of the content. One can post news articles on Twitter and Facebook, create videos on YouTube, and write articles or essays on one's blog. Individuals, not an organization, decide what is important and what they believe should be given the most attention.

  Examples: Blogs, Facebook, YouTube, and Twitter.

Al-Qaeda, the leading force behind most of the current jihadi movements, is today a decentralized network of networks with no structure, hierarchy, or center of gravity. It is based on a global alliance of autonomous groups and organizations in a loosely knit international network. This composition is strikingly similar to the Internet, with its unstructured network and reliance upon a decentralized web of nodes with no center or hierarchy. The parallel between the two may not be so coincidental: al-Qaeda adopted the Internet and has become increasingly reliant on it for its operations and in order to survive. The war on terrorism destroyed al-Qaeda's sanctuary in Afghanistan and forced the organization to transform into a highly decentralized network of alliances and confederation of affiliates. For the new global network of al-Qaeda, the Internet became a crucial platform, carrier, and bonding mechanism (Weimann 2008a). As the most contemporary media outlet, it has become the leading instrument of al-Qaeda's communication, propaganda, recruitment, and networking. Al-Qaeda operatives and supporters are operating numerous websites, forums, chatrooms, Twitter, Facebook, YouTube, and MySpace accounts, with more appearing each year.

Al-Qaeda Goes Virtual

Al-Qaeda's official online debut dates back to February 2000, with the creation of maalemaljihad.com. This was followed in March 2001 by alneda. com, which was active through mid-July 2002. (*"Al-neda"* means "the call" in Arabic.) It was registered in Singapore and appeared on Web servers in Malaysia and Texas before it was taken down at the request of US officials. It then changed its name and URL every few days, forced to move from server to server by citizens who complained to the Internet service providers (ISPs) that were hosting the sites. Then, in July 2002, al-Qaeda lost its Internet domain: it expired and was acquired by a private citizen. The alneda.com site operators tried to reappear by using various server accounts that had no associated domain name. When that failed, they started posting the alneda.com site as a "parasite": the site would be posted on a hijacked website until someone noticed and convinced the ISP to remove it. When it was removed, the process started all over again. In April 2003, al-Qaeda's website reemerged, this time named "Faroq," and it waved the banner of alneda.com. Although the new site and other al-Qaeda sites moved regularly, various informal means were used to pass on details of the site's new locations, including email correspondence, chat rooms, and announcements or links on other groups' websites (Weimann 2008a).

In the summer of 2001, al-Qaeda created its media arm, the As-Sahab Foundation for Islamic Media Publication, and released its first video, "The Destruction of the American Destroyer [USS] Cole." Several other websites at the time were not directly connected with al-Qaeda, but sympathized with its jihadi vision. Websites of this type include Azzam Publications, At-Tibyan Publications (which had one of the earliest jihadi-leaning English-language interactive forums), and Sawt al-Qawqaz. This top-down phase allowed al-Qaeda to control who produced and disseminated jihadi materials online. However, the emergence of the Web 2.0 interactive approach to site design and the growing popularity of interactive forums in the mid-2000s shattered the elitist nature of jihadi communications. Social media platforms such as blogs, Facebook, YouTube, and Twitter enabled global jihadists to share news items, original articles and essays, books, videos, and personal accounts. As Aaron Zelin notes (2013, 5), "The newer technologies lowered the bar for participation, making the involvement of low-level or non-jihadis in the online conversation a new feature of the global jihadi movement."

Today, since al-Qaeda is on the run, its organization is even more virtual, which means greater dependence on the Internet to spread propaganda and plot operations. This reliance on the free access and use of the Internet is also one of the main reasons why, despite the many blows that it has received since 9/11, the organization's operational capabilities have not truly diminished. The Internet is becoming a major weapon in al-Qaeda's bid to win supporters to its cause, preserve its decentralized structure, galvanize its members to action, and raise funds. A widespread network of online platforms is used to feed directions and information from the group's top leadership to supporters and sympathizers around the world and among the jihadists themselves. Al-Qaeda openly acknowledges the importance of the Internet as a propaganda tool, as it did on one of its numerous websites (the Azzam site):

> Due to the advances of modern technology, it is easy to spread news, information, articles and other information over the internet. We strongly urge Muslim internet professionals to spread and disseminate news and information about the jihad through e-mail lists, discussion groups and their own websites. If you fail to do this, and our site closes down before you have done this, we may hold you to account before Allah on the Day of Judgment. . . . We expect our website to be opened and closed continuously. Therefore, we urgently recommend any Muslims that are interested in our material to copy all the articles from our site and disseminate them through their own websites, discussion boards and e-mail lists. This is something that any Muslim can participate in, easily, including sisters. This way, even if our sites are closed down, the material will live on with the Grace of Allah. (Quoted in Anti-Defamation League 2002, 14)

The most visible part of al-Qaeda's online presence involves the spread of propaganda, which is created by the group's media production branch, As-Sahab. This organization uses modern technology and semiprofessional hardware to produce its video statements and distribute them worldwide. In addition to being released in Arabic, some published videos include subtitles in English or other languages, while more recent productions include videos in English and German. Al-Qaeda also operates radio and television broadcasting online along with its online production facility—the Global Islamic Media Front, one of al-Qaeda's mouthpiece groups (Weimann 2008a).

In testimony before the US House of Representatives Homeland Security Subcommittee on Counterterrorism and Intelligence, Brian Jenkins from the RAND Corporation presented the report *Is Al Qaeda's Internet Strategy*

Working? (Jenkins 2011) He described al-Qaeda as the cutting edge among terrorist groups in cyber know-how: "While almost all terrorist organizations have websites, al Qaeda is the first to fully exploit the Internet. This reflects al Qaeda's unique characteristics. It regards itself as a global movement and therefore depends on a global communications network to reach its perceived constituents. It sees its mission as not simply creating terror among its foes but awakening the Muslim community. Its leaders view communications as 90 percent of the struggle" (Jenkins 2011, 1). According to the RAND study, al-Qaeda's websites fall into three categories. At the top are the official sites that carry messages of the leaders. Recognized jihadist figures discuss issues of strategy on a second tier. The third tier comprises the many chatrooms and independent websites where sympathizers and followers verbally and visually embellish the official communications; fantasize about ambitious operations; and boast, threaten, and exhort each other to action. According to Jenkins (2011, 2), "The quantity and easy accessibility of these sites have attracted a host of online jihadists, some of whom are technically savvy and contribute their skills to the overall communications effort."

The unique unstructured and decentralized nature of al-Qaeda affects its online production line, or, in fact, production *lines*: Al-Qaeda is based in numerous branches, with the leading ones being Al-Qaeda in the Arabian Peninsula (AQAP), Al-Qaeda in Iraq, and Al-Qaeda in the Islamic Maghreb (AQIM). Each of these branches has its own media outlet—in Pakistan, al-Qaeda has As-Sahab Media; in Yemen, Al-Malahim; in Iraq, Al-Furqan; and in North Africa, Al-Andalus. Each of these production lines involves a multistep process: first, they produce the videos or statements from these organizations or their leaders. Then they edit the material, inserting sound effects, pictures, or video clips, and adding translations or subtitles. Finally, they send the finished product(s) off to their media outlets, which function as a distribution network online. Often, the administrators of jihadi forums or social media users will take the content and post it onto online threads and thus promote its exposure. The motive for al-Qaeda's reliance on leading online forums is authentication: By distributing its messages through accredited venues (such as the forums Shamukh al-Islam and al-Fida al-Islam), it assures viewers that the information is truly an official statement. As a result, when the Shamukh al-Islam and al-Fida al-Islam forums went down in March 2012, neither al-Qaeda nor its affiliates distributed any new products. Nothing was officially released from al-Qaeda until Shamukh al-Islam came back online. However, within six hours of Shamukh al-Islam's return online, a new release from the Global Islamic Media Front had been posted, and within nine hours there was a new release from al-Fajr (Zelin 2013).

Finally, al-Qaeda has undergone a structural shift from being a strictly centralized, hierarchical organization to a huge global network of affiliated, semi-independent cells that have no single commanding hierarchy. The use of the Internet allows these loosely interconnected networks to function, communicate, and maintain their ideological solidarity. The Internet connects not only members of the "hardcore al-Qaeda," but also members of numerous groups who associate themselves with the jihadist spirit. For instance, the websites of many jihadist groups affiliated with al-Qaeda are interlinked, expressing support and solidarity. These sites and their related forums permit terrorists in places such as Afghanistan, Chechnya, Indonesia, Iraq, Lebanon, Malaysia, Palestine, the Philippines, and Turkey to not only exchange spiritual and religious messages but also provide online training and practical information about terrorist tactics and practices.

Recently, as the 2011 RAND report highlights, al-Qaeda has embraced individual jihad as opposed to organizationally led jihad: "Increasingly, it has emphasized do-it-yourself terrorism. Those inspired by al Qaeda's message are exhorted to do whatever they can wherever they are. This represents a fundamental shift in strategy. As part of this new strategy, al Qaeda has recognized online Jihadism as a contribution to the jihadist campaign" (Jenkins 2011, 4). This shift also appears in the online platforms used by al-Qaeda: there is a steady transition to new channels such as social media platforms that target individuals, emphasize interactivity, and appeal to youth.

Al-Qaeda on Social Media

In 2003, the prominent al-Qaeda strategist Abu Musab al-Suri posted *The Call to Global Islamic Resistance*, a massive document outlining a series of suggestions for the future of jihad. In it, he stressed the need for al-Qaeda to restructure itself according to the notion of "*nizam, la tanzim*" ("system, not organization"), an evolution toward the highly decentralized, multinodal network that defines al-Qaeda today. Around the same time, the term "Web 2.0" was coined, to describe the cyber-enabled, peer-to-peer network shape that the Internet began to take. Among the many manifestations of Web 2.0 was the emergence of new media, a growing array of interactive communications systems facilitated by a rapidly expanding set of platforms. These included blogs, Web forums, Facebook, MySpace, Twitter, and YouTube, all linked together in innovative ways that helped begin to form the new media landscape. John Curtis Amble (2012, 339) noted that "[t]he similarities between these two structural transformations, one of a transnational terrorist group and the other of the Internet, are striking. Indeed, Al Qaeda seems to have acknowledged these similarities by increasingly operating with considerable effectiveness in the new media environment."

Since the 1990s, jihadi terrorists have leveraged the power of the Internet in more imaginative ways: almost all jihadi are active, both publicly and discreetly, on social media networks such as Facebook, Twitter, MySpace, YouTube, Vimeo and Instagram (Weimann 2010a, 2010b; Weimann and Gorder 2009; Stalinksy 2013b). As noted by terrorism expert Evan Kohlmann, the trend is clear: "Today, 90% of terrorist activity on the Internet takes place using social networking tools, be it independent bulletin boards, Paltalk, or Yahoo! eGroups. These forums act as a virtual firewall to help safeguard the identities of those who participate; they offer subscribers a chance to make direct contact with terrorist representatives, ask questions, and contribute and help out the cyberjihad" (quoted in Noguchi 2006). As new generations have come of age in the Internet era, jihadi groups (especially those associated with al-Qaeda) have spread their online presence, establishing what terrorism expert Evan Kohlmann (2011, 3) described as "a tenacious beachhead in cyberspace . . . a cabal of critical Jihadi-oriented online social networking forums." As Amble (2012, 339) reports, "Terrorist groups are now harnessing the unique characteristics of the new media environment that has taken shape in the past decade, while security services struggle to conceptualize this rapidly evolving virtual landscape."

Numerous al-Qaeda groups and affiliates are increasingly using social media: The al-Qaeda branch in Yemen has proved especially adept at disseminating teachings and commentary through several different social media networks. The Jabhat al-Nusra (Al-Nusra Front), an influential jihadist group in Syria, tweets mostly in Arabic to its followers, but there are sparing tweets in English that are clearly trying to appeal to a wide swath of potential recruits and thought leaders from the global community. Groups like AQIM have established robust Twitter accounts and social media strategies to push rhetoric and propaganda from North Africa to broader audiences around the world. A 2009 report compiled by the Simon Wiesenthal Center found a number of terrorist-related Facebook groups, including one called "HAMAS Fans" and another titled "Support Bin-Laden (Al-Qaeda)—Eradicate the West!!!!" (Cooper 2009). Somalia's al-Qaeda affiliate, al-Shabaab, is microblogging to get its message out: "Our war against the West is a war for the sovereignty and dominance of Allah's Law above all creation. No to democracy and Kafir laws!" it tweeted through its now-suspended Twitter account, using the Arabic word for "infidel" to spread its propaganda (quoted in Clayton 2013).

These interactive channels are clearly having some impact on some of the participants. According to the report presented by Kohlmann (2011), each week new Internet personalities disappear from the Web on a mission to live out their outlandish jihadi fantasies. Kohlmann noted that at least 120 such individuals (including US nationals) have graduated from being mere "pajama-hideen" to taking a real role in terrorist activity over the past seven years. Of these 120 hardcore extremists, more than half were killed. An illustrative example is that of Ansar al-Mujahideen forum user "Khattab 76," who was reported dead on August 23, 2011, after clashes with the Egyptian military in the Sinai Peninsula, where he had gone to "fight the Zionists." According to Ansar forum administrators, inspired by what he saw on the Web, "Khattab 76" had made several previous failed efforts to join al-Qaeda in both Iraq and Afghanistan. Another case involved a participant in al-Qaeda's Shamukh forum: On September 18, 2011, moderators on this forum informed their comrades that user "Qutaiba" had departed for Algeria to join AQIM. They quoted a final message from him sent over the Internet: "I am here amongst the mujahideen in the Islamic Maghreb . . . I advise my beloved ones to join the convoy before it is too late . . . Your brothers in AQIM are waiting for you."

The most famous individual to self-recruit on the Internet using al-Qaeda's online social networking was a young Jordanian doctor named Humam al-Balawi (also known as Abu Dujanah al-Khorasani). On December 30, 2009, Balawi, a former administrator on al-Qaeda forums, blew himself up at a secret CIA base along the Afghan-Pakistani border, killing seven CIA officers and one Jordanian. At the time, CIA intelligence agents believed they had successfully recruited Balawi as a double agent to help hunt down Ayman al-Zawahiri and other top al-Qaeda figures. In fact, Balawi was offering a starkly different perspective to his associates on the jihadi web forums. In an interview published on al-Qaeda's Al-Hesbah forum in September 2009, only three months previous, Balawi appealed, "How can I encourage people to join the jihad while I'm staying away from it? . . . How do I become a burning wick for others follow the light of? Can any sane person accept that? Not me." He also noted the role of al-Qaeda's social networks: "I left behind on the forums some brothers who are dearer to me than members of my own family. . . . When I meet any mujahid here who knows about the forums, I rush to ask him who he knows from al-Hesbah—as he might be one of those whom we loved in the cause of Allah, from amongst the administrators or members, and I would hug him as one brother longing for another."[7]

In late 2013, two Palestinian men, one in Gaza and one in east Jerusalem, allegedly conducted Internet conversations for the purpose of planning a series of terror attacks. On one end of the chat, in Gaza, sat a man named Ariv al-Sham, who took his orders directly from the head of al-Qaeda central, Ayman al-Zawahiri. On the other end was a 23-year-old resident of east Jerusalem, Khalil Abu Sara. The targets chosen by the two were the Jerusalem Convention Center and the US Embassy in Tel Aviv. The entire communication took place in chats over Skype and Facebook. The two men settled on a plan to bomb both sites simultaneously, and to murder emergency responders in Jerusalem with a follow-up suicide truck bomb. In December 2013, they were arrested, and during his questioning Abu Sara told the Israeli interrogator that his handler in Gaza made it clear to him that he was in direct online communication with Zawahiri. How did the two operatives find each other? According to the investigation, Abu Sara first trawled the Internet, asking Salafi jihadi elements online about where he could get hold of explosives. Abu Sara used the social net to speak to bomb experts from Gaza, learning about the process of assembling explosives. At some point, in the maze of the online jihadi world, he "met" Ariv al-Sham (Lappin 2014).

What may be most alarming about this phenomenon is the sharp increase in the use of brand-name US commercial social networking services such as YouTube, Twitter, and Facebook by terrorist organizations and their supporters. The emerging trend presents new challenges to counterterrorism officials and experts working to quell and prevent terrorist actions. There is no doubt that YouTube and Facebook have been making genuine efforts in an attempt to thwart the online activities of al-Qaeda supporters and operatives. However, a simple search for jihadi videos on YouTube will reveal hundreds of al-Qaeda video clips, a compelling demonstration that these efforts have thus far been insufficient in addressing the problem.

Rising Interest in the Field

The combination of terrorism cyberspace, the emergence of cybersavvy terrorists, and the growing reliance of modern terrorism on cyber tools and

---

[7] Quoted in the September 2009 *Vanguards of Khorasan* online magazine 1 (15).

platforms has yielded increasing interest levels from both academia and the media. It is useful to present some of the indicators of this growing interest as well as some findings from analyzing these indicators.

The Growing Scholarly Interest

One quick glance through a search engine confirms that terrorism on the Internet has become a "hot" issue: A Google search combining the keywords "terrorism" and "Internet" yields over 109 million results. How many of these results are based on academic, scholarly interest? To assess this, one must turn to scientific publications in this field. Two separate and leading academic databases, EBSCO Information Services and PAIS International, were used to search for relevant publications. EBSCO provides full-text databases, subject indexes, historical digital archives, and e-books. Within more than 375 databases, it has tens of thousands of full-text journals and more than 350,000 e-books. PAIS contains citations for journal articles, books, government documents, statistical directories, research reports, conference reports, publications of international agencies, microfiche, Internet material, and more, from more than 120 countries.

The EBSCO and PAIS search period was conducted from 1996 (when, after the 1995 Oklahoma City bombing, the public first became aware of the existence of "Mayhem Manuals"—online guidebooks on explosives, poisons and terrorism) to the present time of writing (2014). After the two searches were compared, the results were pooled in a comprehensive literature list. Scientific terrorism research was separated from publications concerned with information technology security and cybercrime. Therefore, a necessary requirement was that every publication should present analysis and/or findings regarding online terrorism and not just speculations on potential cyberattacks or information technology security shortcomings that could facilitate such attacks. The searched sources were limited to scholarly publications, namely journal articles, reports, books, conference papers, and published theses (if archived). Subsequently, the publications' relevance was assessed using four specific criteria: relevancy, innovation, reliability, and length.[8]

---

[8] Each item was rated on a scale of 0 to 3 points on all four criteria. The score for the first two criteria was attributed based on the information given in the publication's abstract, and then multiplied by either four (relevancy, in terms of the extent to which terrorist behavior was analyzed) or two (innovation, in terms of newly collected dates or newly postulated/applied theories). The reliability score was based on the source: peer-reviewed academic publications scored 3 points,

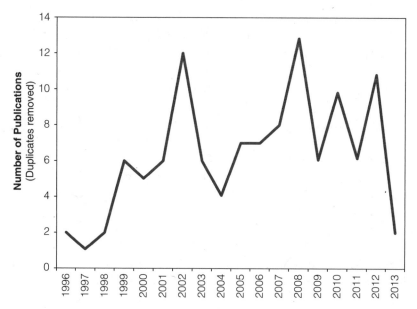

Figure 1.1. Academic Publications (EBSCO and PAIS) on Internet Terrorism, 1996–2014

Sources: EBSCO Information Services and PAIS International

The EBSCO and PAIS search results—94 and 114 hits, respectively—were combined to an overall list of 150 relevant publications. Of the EBSCO searches, 34.4 percent were deemed to be highly relevant, 46.8 percent were deemed relevant, and 19.1 percent were deemed less relevant. Of the PAIS searches, 28.1 percent of the results were assessed as highly relevant, whereas 41.2 percent were labeled relevant and 30.7 percent were labeled less relevant. As figure 1.1 reveals, there is a gradual increase in publications after the 9/11 attacks, although this is attributed mainly to the extensive period of writing and research that manifested mostly in 2002. Furthermore, the searches show a varying but on average increasing output of publications with an interesting "peak" in 2008—which may be explained by several

---

non-peer-reviewed academic publications scored 2, peer-reviewed nonacademic publications scored 1, and non-peer-reviewed nonacademic publications scored 0 points. This score was also multiplied by two. The score for length was based on an assumed average for a publication (e.g., 15 pages for an article, 150 pages for a book). Publications that scored from 0 to 15 points overall were rated as less relevant, those that scored between 16 and 20 points were classified as relevant, and those that scored more than 21 points were deemed highly relevant.

major events in that year, including the Mumbai attacks, which consisted of 12 coordinated shooting and bombing attacks carried out over four days in late November by members of the South Asian terrorist group Lashkar-e-Taiba, resulting in 164 deaths and more than 300 injuries. The apparent decline in 2013 is a spurious event, caused by the delay in the databases' recording of recent publications.

Rather surprisingly, both searches reveal that more than 84 percent of the authors wrote only one journal article on the subject. In fact, only three authors published five or more articles,[9] 9.8 percent of the authors wrote two articles, 2.3 percent wrote three articles, and one author (0.8%) wrote four articles. Undoubtedly, scholarly interest in the subject has increased. However, there are a surprisingly limited number of scholars who consistently study this subject. Most of the publications come from a one-time exploration or "temporal" visit. This finding may imply that the scholarly discourse in this field has to be fostered and shaped in order to keep up with the recent developments and provide valid results that may have counter-terrorism applications.

The Growing Media Interest

Terrorist events are major news items, and numerous studies have documented the media's almost obsessive interest in terrorism (e.g., Nacos 2002, Weimann and Winn 1994). Regarding online terrorism, journalists occasionally write about "terrorist use of the Internet" (for instance, the infamous beheading of US citizen Nicholas Berg by Iraqi insurgents, which was videotaped and later broadcast online). When trying to examine the media interest in this topic, one encounters several problems. First, as expected in similar studies of scholarly discourse, almost every keyword-based search on this subject brings up undesired hits from the fields of cybersecurity, cybercrime, and cyberwarfare. Unfortunately, a keyword-based search appeared to be the only possible procedure, since the amount of archived articles is overwhelming. Second, a comprehensive analysis of an entire national, let alone global, coverage for the study period (1996 to 2014) is extremely demanding or even impossible, owing to the huge amount of items and the lack of a global media database. Therefore, our search was limited to the two leading US newspapers, the *Washington Post* and the *New York Times*. These two major papers were selected as representing the print and online mainstream

---

[9] These are Maura Conway, Jonathan Matusitz, and Gabriel Weimann.

media; they were also selected because they are considered to be leading the world media, and both are often quoted worldwide.

The online archives of both newspapers were scanned with the sophisticated search engine NEXUS, using the combination of appropriate keywords relating terrorism and the Internet or cyberspace. To focus on analytic articles rather than on news reports, the articles had to be at least 800 words long. Furthermore, to exclude undesired results dealing with the setup of cyberdefense and cyberwarfare capabilities, any article with the acronym "NSA" (for the US National Security Agency) in its headline was dismissed. This procedure obviously has some weaknesses and is by no means guaranteed to find all relevant items or list only relevant articles. However, after conducting sample checks, the procedure seemed to sufficiently serve the purpose. Figure 1.2 presents the results of the search.

The scan revealed 548 relevant articles, of which 241 were published in the *Washington Post* and 307 were in the *New York Times*. The distribution over time shows the rising coverage of this issue after 9/11. The media

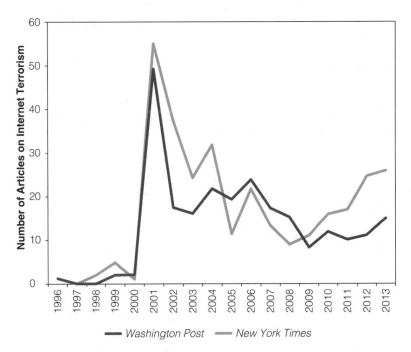

Figure 1.2.   Journalism (*New York Times* and *Washington Post*) on Internet Terrorism, 1996–2014

interest then decreased, reaching its lowest point in 2008, as shown in figure 1.2. However, since then, the coverage has increased to around 40 articles per year. This rise may reflect the impact of the numerous terrorist actions in 2008, including the Mumbai deadly attacks. The search also revealed that 69.7 percent of the journalists wrote only one of the articles, 24.5 percent wrote two to four articles, 2.6 percent wrote five to seven articles, and 2.0 percent wrote eight to ten. Only 4 out of 343 authors (1.2 percent) wrote more than 10 articles.[10]

Although the interest in the subject apparently has increased since 2008, it should be noted that this development does not represent the growing importance of the topic. Ironically, quite the opposite is the case: after 9/11, many terrorist safe havens and camps were destroyed, and subsequently the importance of the cyberspace and online platforms grew dramatically. Moreover, it appears, again, that this topic is covered mostly by occasional contributors (70 percent write only one item), with very few specialized journalists who consistently report on or analyze the topic of online terrorism.

Terrorism entered cyberspace during the 1990s and has stayed there ever since. All terrorist groups are today on the Internet, using freely the most advanced platforms of online communication. Though terrorists never developed or invented new online technologies, they were very quick in learning and applying the newest forms of these online channels. Consequently, their online presence has changed considerably over the years. As the next part reveals, the combined impact of technological changes, growing sophistication by cybersavvy terrorists, and the launching of online counterterrorism measures was a dramatic change in terrorist cyber-presence. The following chapters describe these emerging trends in online terrorism.

---

[10] These are Neil MacFarquhar, Robert Mackey, John Schwartz, and Scott Shane.

Part II

Emerging Trends

Terrorist presence in cyberspace has changed considerably. When I completed my book *Terror on the Internet* (Weimann 2006b), I assumed that the new virtual arena would change rapidly. And yet, I did not realize then how dramatic and multifaceted this change would be. In fact, in the years following the book's publication I noted that although some of the predictions I made were accurate, many of them failed to estimate the magnitude, sophistication, and volume of terrorist use of cyber platforms. The combined effect of technology, online terrorism, and counterterrorism resulted in a powerful dialectic process that changed the terrorist cyber-presence dramatically. In this ever-changing, dynamic, and restless environment, several trends have emerged. Some of them emerged recently, and some have only recently become more significant and dominant. The following chapters will describe some of these trends, as documented through continuous monitoring and analysis of terrorist use of the Internet.

Chapter 2

Narrowcasting

An emerging trend in online terrorism is "narrowcasting": the dissemination of information (usually through radio, television, or the Internet) to a narrow audience, not to the broader public at large. Also called "niche marketing" or "target marketing," narrowcasting aims media messages at specific segments of the public, defined by characteristics such as values, preferences, demographic attributes, or location. The concept is based on the postmodern idea that mass audiences do not exist (Goncalves, Kostakos, and Venkatanathan 2013). Terrorists have learned about this new concept and now apply it in their cyber-campaigns. Instead of "one website for all," Internet-savvy terrorists target specific subpopulations, including children, women, "lone wolves," overseas communities, and diasporas. Several examples will illustrate this trend.

This chapter updates and expands on material originally published in "How Terrorists Use the Internet to Target Children" (Weimann 2008c), "Online Terrorists Prey on the Vulnerable" (Weimann 2008d), and "Virtual Sisters: How Terrorists Target Women Online" (Weimann 2009b).

Targeting Children Online

Terrorists recognize the Internet's popularity among children and youth, and increasingly use it to target them. One of Hamas's websites, the online magazine *Al-Fateh* ("The Conqueror"), is designed for children, with cartoon-style design and colorful children's stories. The site's title promises "pages discussing Jihad, scientific pages, the best stories (that cannot be found elsewhere), and unequalled tales of heroism." The *Al-Fateh* site, which is updated regularly, has a link to the official Hamas site, www.palestine-info.com. Mixed among attractive graphics, children's songs, and stories written by children themselves are messages promoting suicide terrorism. Thus, the site presented a gruesome photo of the decapitated head of young Zaynab Abu Salem, a female suicide bomber who, on September 22, 2004, detonated an explosive belt in Jerusalem, killing two policemen and wounding 17 civilians. The caption for the image praises the act, arguing that she is now in paradise, a "*shaheeda*" (martyr) like her male comrades: "The perpetrator of the suicide bombing attack, Zaynab Abu Salem. Her head was severed from her pure body and her headscarf remained to decorate [her face]. Your place is in heaven in the upper skies, O Zaynab, sister [raised to the status of heroic] men." The same website posted the last will of a Hamas suicide bomber, who, on June 1, 2001, carried out a suicide bombing attack at the Dolphinarium, a teen club in Tel Aviv. The attack resulted in the deaths of 21 Israeli civilians, most of them teenagers. In his online will, the suicide bomber wrote: "[T]he true heroes are those who write the history of their people in blood . . . I will turn my body into shrapnel and bombs, which will chase the children of Zion, will blow them up and will burn what is left of them . . . There is nothing greater than killing oneself on the land of Palestine, for the sake of Allah."

Comic strips are another online apparatus in terrorist seduction of children. Jihadists produced a comic strip called "Son of the Martyr," and posted it on the Shumukh al-Islam forum on September 30, 2011. The first issue of the comic presents the story of three brothers, the sons of a slain fighter, who use an explosives-laden toy car to blow up a tank of enemy soldiers. In the comic, the youngest brother Qassim is approached by soldiers and given a remote-controlled toy Hummer in return for acting as a spy. His brothers Abdul Qadir and Saif decide to use the toy against the soldiers. As Qassim sends the explosives-laden toy toward the target, he says, "Now, I will avenge and retaliate for my religion, my honor, and my land."

Computer games are a further tactic for targeting children online. Several terrorist groups offer free online games designed as instruments of radicalization and training. One such free online game was the "Quest for Bush," also known as "Night of Bush Capturing," which was released by the Global Islamic Media Front, an al-Qaeda media outlet. Armed with a rifle, a shotgun, or a grenade launcher, players navigate various missions that include "Jihad Growing Up," "Americans' Hell," and "Bush Hunted Like a Rat." In the final stage, the player's task is to kill President Bush. On March 22, 2011, an al-Qaeda-themed modification for the first-person multiplayer computer game shooter "Counter-Strike" was launched on the Shumukh al-Islam forum. It was titled "Alqaida-Strike 1.4," and the options allowed a player to use the banner of the al-Qaeda-affiliated Islamic State of Iraq (ISI). During gameplay, jihadi chants that are usually heard in video productions by al-Qaeda and its affiliates are played in the background.

Another example of this type of online gaming is Hezbollah's "Special Force," which allows players to become warriors in a terrorist campaign against Israel. The violent game features a training mode in which participants practice shooting skills on former Israeli prime minister Ariel Sharon and other Israeli political and military figures. A high score in the game earns a special certificate signed by Hezbollah leader Hassan Nasrallah and presented in a "cyberceremony." At the end of the game, the players receive a special display of Hezbollah "martyrs"—fighters killed by Israel. In August 2007, a new version emerged, entitled Special Force 2. Based on the 2006 Lebanon War between Hezbollah and Israel, and available in Arabic, English, French, and Persian, the game allows players to take on the role of a Hezbollah combatant in a 3D environment.

In 2012, jihadists posted several games supporting Muslim fighters in Chechnya, Gaza, Iraq, Mali, and Tunisia. These games enable players to become fighters to protect homes, rescue prisoners, and blow up enemy aircraft. The first two games were uploaded to the Archive.org website on September 17, 2012, and they were advertised on the Ansar al-Mujahideen forum. In the first game, "Gaza Strip," a player pilots a plane to shoot Israeli aircraft as a speech from former al-Qaeda leader Osama bin Laden plays in the background. When the player is hit 10 times, a message is displayed, reading: "Congratulations, you have been martyred." The second game, "Caucasus," contains the same bin Laden speech. The player maneuvers a rocket launcher to shoot Russian helicopters. Text displayed before the game reads: "Kill the Russians, down their planes, rub their noses in the dirt, and avenge our brothers and sisters in the wounded Caucasus." The third game, "Female Prisoners," was uploaded on December 4, and has the player control a fighter to pass obstacles to reach a female prisoner. The fourth game, "Support Our Brothers and Sisters in Tunisia," was uploaded on January 8, 2013; in it, the player protects a house from bombs dropped by the Interior Ministry of Tunisia's ruling Ennahda government. The fifth and latest game was uploaded on February 4, 2013. Titled "Muslim Mali," the game allows users to pilot a plane and shoot down French aircraft. In the message posted about the game on February 14, 2013, the jihadist poster "Ta'ir al-Nawras 07" encouraged fellow jihadists to learn how to create similar games using the Construct 2 game-designing software.

In March 2013, Al-Qaeda in the Islamic Maghreb (AQIM) launched an online campaign targeting young children to lure them to its radical ideology. To do this, the AQIM-produced video games include a clear strategy that showcases the group's ability to win wars against international forces. One game, designed and posted online by AQIM, has a military aircraft carrying AQIM's black flag while hitting and destroying French aircraft in the Sahara, where battles are raging against the terrorists in northern Mali.

Numerous studies have assessed the impact of violent computer games and video games on children: they reveal the effects in terms of desensitization to the use of violence and growing likelihood of using violence: Research on exposure to television and movie violence suggests that playing violent video games will increase aggressive behavior. A meta-analytic review of the video-game research literature reveals that violent video games increase aggressive behavior in children and young adults (Anderson and Bushman 2001). Experimental and nonexperimental studies with males and females in laboratory and field settings support this conclusion. Analyses also reveal that exposure to violent video games increases physiological arousal and aggression-related thoughts and feelings. Playing violent video games also decreases prosocial behavior. One may deduct that the terrorist games, especially designed to achieve such impact, may have even more negative effects.

In September 2012, a Facebook page for Fatah in Lebanon posted a picture of a mother dressing her young son with a suicide belt. This picture was posted on the Fatah site together with an imaginary conversation between the son, who is being sent to his death, and the mother, encouraging it. "Why me and not you?" the child innocently asks his mother, who answers that she will continue to have more children "for the sake of Palestine" (Marcus and Zilberdik 2012). The Internet is also used to recruit and train children: in February 2008, the American army in Iraq revealed a video captured in a raid in Diyalla province. The video, which had been posted on the Web, depicts children in ski masks kidnapping a grown man on a bicycle, sitting in a circle surrounded with firearms and singing al-Qaeda songs, and storming a room with bound adult hostages and waving guns at their heads. The children are not actors recruited solely for the sake of propaganda, but are in fact active participants in the group's violent activities (Holden 2008). Jonathan Evans, the head of the British domestic intelligence service MI5, said in a 2007 interview that al-Qaeda was recruiting Muslim children as young as 15 to wage "a deliberate campaign of terror," and warned that Islamists were "radicalizing, indoctrinating and grooming young, vulnerable people to carry out acts of terrorism" (Johnston 2007). Clearly, the Internet has become one of the most efficient instruments for targeting these young audiences.

Targeting Women Online

She Penetrated Deeper into the Crowd, Her Breath Coming in Spurts and the Spark of Death Shining in Her Eyes, and . . . Allah Akbar! . . . The Body-Parts of the Jews Scattered Everywhere, and the Black Blood Spilled on the Soil of the Prophets, While Her Pure Soul Ascended to Paradise.

—From a posting on the al-Hesbah forum, July 2007

The ninth issue of *Sawt al-Jihad* (Voice of Jihad), an al-Qaeda online magazine, was one of the first in a series of female-oriented postings. Published in January 2004, the issue included a special section for women and attempted to recruit women for terrorist attacks. One article, "Um Hamza, an Example for the Woman Holy Warrior," is the story of a female martyr as told by her husband. Later, an online women's magazine called *al-Khansa*— named after a seventh-century female Islamic poet who wrote eulogies for Muslims who had died while fighting the "infidels"[1]—was launched. The website gives women advice on raising children to carry on the jihad, using first aid to treat family members injured in combat, and preparing to fight

---

[1] Al-Khansa is also called the "Mother of Martyrs," because after her four children died in one of the battles of early Islam she did not mourn them, but rather thanked Allah for "honoring her with their deaths."

through physical training. The main goal of the magazine seems to be to teach the wives of terrorists how to support their husbands in the violent war against the non-Muslim world. One of the magazine's first articles reads: "The blood of our husbands and the body parts of our children are our sacrificial offering."

Even the wife of al-Qaeda leader Ayman al-Zawahiri is used for female-oriented online campaigns. In June 2012, Al-Fajr Media Center released a message from Umayma Hassan, al-Zawahiri's wife, addressing Muslim women in the countries that experienced the 2011 Arab Spring. The message, titled "To the Muslim Women after the Revolutions," was posted on jihadist forums. Hassan told Muslim women to remain faithful to Islam and observe its traditions such as wearing the hijab (headscarf). Later that year, the "Workshop of the Granddaughters of Safiya," a female-focused jihadist media group, lauded Hassan for her advice and asked that Muslim women be strong in supporting Islam and assisting the male fighters. The group wrote in an online posting: "By Allah, it is now the time when it is impermissible to stay behind and watch the events unfold from a distance, while everyone is seeing the division of the world into two sides: the side of the truth and the side of the falsehood. Don't hold back your help and aid from your brothers" (quoted in SITE Monitoring Service 2014a).

A number of jihadist websites include special subforums for women: www.al-hesbah.org has a subforum for women and Islamic families; www.shmo5alislam.net has a women's subforum; and www.al-faloja.info/vb has a family subforum. A review of these forums reveals that their objective is essentially indoctrination, and that their principal concern is with encouraging women to carry out suicide attacks—for example, by posting biographies and testaments of female martyrs in both Islamic history and modern times. They also promote jihad by citing various fatwas (religious rulings) on jihad and martyrdom, and they urge women to take an active part in jihad, or at least to support its fighters. In the al-Hesbah forum, a writer named Umm Hamza al-Shahid posted a message in July 2007 titled, "Secure Yourself a Chandelier under the Throne [in Paradise]," in which she encouraged Muslim women to carry out suicide bombings (Hazan 2008).

A series of messages regarding Muslim women and their role in supporting jihad and its war on the West have also emerged on jihadist forums. One document, titled "What Do the Mujahideen Want from a Muslim Woman?" was written by Abu Omar Abdul Bar. Bar charges that the war launched by the West against Islam is not limited to direct military action, but also seeks to affect economic, social, and character changes. In this regard, he believes that the enemy is trying to alienate Muslim women within their community and provoke them against Muslim society. Contrary to the alleged Western desire to make women a "cheap commodity," this message advocates the role of women as mujahideen, citing various examples of female mujahideen in Muslim history. Bar writes that a Muslim woman should support the mujahideen, feed her sons "gunpowder with milk," and raise them in the spirit of jihad.

The use of female suicide bombers by Islamist militant groups in such countries as Afghanistan, Iraq, Israel, Jordan, Sri Lanka, Turkey, and Indian Kashmir is on the rise. It is related to a significant increase in the number of websites dedicated to mujahidat (female mujahideen), martyrdom by women, or the use of women as suicide bombers (see Hazan 2008). In 2004, a woman from Gaza named Rim al-Riyashi became the first female suicide bomber directed by Hamas. She killed four Israelis and wounded seven at the Erez crossing between Gaza and Israel after detonating the bomb hidden under her clothes. She became an icon in Hamas propaganda. In March 2007, a macabre music video appeared on a television show for Palestinian children and on Hamas websites, echoing the story of Rim al-Riyashi. A posting on the jihadist forum al-Hesbah in December 2007, titled "The Islamic Woman Knight," called on Muslim women to follow the example of the women of early Islam and sacrifice their souls for the sake of religion. The text includes the following calls:

> "My sisters, the Muslim jihad fighters. . . . We want to follow the path of grace and jihad taken by our mothers and sisters of the early Islam. . . . Why shouldn't contemporary women jihad fighters . . . sacrifice what is most precious to them, and [give] their souls for the sake of religion[?] . . . Didn't Muslim women in the days of the Prophet and the Righteous Caliphs join the army and set out on jihad?"

There are online rewarding mechanisms for female suicide bombers. For example, the Islamic Movement of Uzbekistan released a video regarding a female suicide bomber named Ummu Usman, who carried out an attack in Pakistan in 2012. On July 9, 2013, a 20-minute Uzbek-language video produced by the group's Jundullah Studio was posted on YouTube; on July 10, it was posted on the Jamia Hafsa Urdu Forum. Footage in the video shows an explosives-laden vehicle driven by Ummu Usman, and the blast and aftermath of her attack; the footage also shows Usman delivering a speech in Russian about women participating in jihad (even though many men do not) and her emigrating from Russia. In March 2013, Maryam Farahat, who was known as Umm Nidal, died. Her six sons were all jihad fighters or martyrs in the Al-Qassam Brigades, the militant arm of Hamas, and three of them were killed. Umm Nidal has been dubbed "the Khansa of Palestine," after the above-mentioned female poet Al-Khansa, and she is praised in numerous online postings, including websites and forums of the Hamas and the Muslim Brotherhood. Hamas leader Ismail Haniya is quoted in these postings, saying: "She [Umm Nidal] is the extraordinary woman who gave birth to a son who led the manufacture of the Qassam rockets . . . [that] reached occupied Tel Aviv and Jerusalem. She [also] gave up her son Muhammad for the sake of Allah. He successfully killed Zionist officers in a heroic battle. . . . [Her sons] Nidal, Muhammad and Rawad are martyrs with a single mother. All were active in the Izz ad-Din al-Qassam Brigades and have an honorable jihadi past."

Calling women to take part and promising them heavenly rewards is often accompanied by online teaching and practical instructions. Al-Fajr Media Center, one of the producers of al-Qaeda's propaganda, told female readers in the 2011 issue of its online magazine *al-Shamikha* (The Majestic Woman): "The enemies wish with the greatest desire to remove her from the truth of her religion and the truth of her role, because they know well how the situation will be if the women entered the battleground." The second issue of *al-Shamikha* was posted on jihadist forums on February 3, 2012, and included as a special insert a document by prominent al-Qaeda member Abu Qatada al-Filistini titled "Jihadi Upbringing and Sacrifice." Similar to the first issue, the second issue contains articles on etiquette, first aid, and skin care. Additionally, the second issue contains pieces aimed to incite support for fighters, and calls all Muslims to defend jihad. A new section of the magazine, "The Digital Majestic Woman," provides an introduction to principles of computer and network security (MEMRI Jihad & Terrorism Threat Monitor 2011b, 2012).

The messages targeting women are also referring to their role as spouses and mothers. On March 6, 2008, the al-Qaeda media production house As-Sahab posted an audio message by Sheikh Mustafa Abu Yazid, an al-Qaeda commander in Afghanistan, on the Islamist website al-Ikhlas. In the message, titled "They Lied and Now [It Is Time for] Combat," Abu Yazid calls upon wives to not stand in their husbands' way to enter paradise, saying, "A righteous wife who loves her husband is one who wishes for him to enter Paradise . . . and who says to him . . . 'Take my gold and property and wage jihad for the sake of Allah . . . and we shall meet in Paradise, Allah willing . . .'" The Al-Faloja website posted a story describing the reaction of a mother who had been informed that her four sons had become martyrs: "The [mother] was happy . . . and said: 'Praised be Allah for honoring me with their martyrdom. I pray for Allah to let me join them in His abode. Tell me now, what fate did Allah grant you, [the jihad fighters]?' He replied: 'We attained a decisive victory.'" In another jihadist forum, al-Hesbah, a posting appeals to mothers of mujahideen, expounding the importance of their sons' role as defenders of the faith and detailing the advantages of martyrdom:

> To every mother of a jihad fighter combating Allah's enemies. . . . You have borne us a hero—one of the Islamic heroes of which the nation is proud and who are [fighting on the various jihadi] fronts. . . . Were it not for Allah and for [your son] and his brothers, Muslim lands would have become forfeit, and religious commandments would have been annulled; however, owing to his ideals, [your son] stands like a bastion against the enemies of religion. Hence, [as] his mother, you are blessed. . . . Wouldn't you like your son to attest to your righteousness on Judgment Day? Wouldn't you like his death to be easy? . . . Wouldn't you like your son to be among the best of the martyrs of Allah's Messenger? . . . Wouldn't you like Allah to extol your son before His angels? (al-Hesbah online forum, October 11, 2007).

These attempts at targeting children and women are just two examples of terrorists' increasing use of online narrowcasting. They use this tactic to appeal to, seduce, and recruit targeted subpopulations, including members of the so-called diaspora communities or potential supporters living overseas, in Western societies. The success of the Islamic State and other jihadi groups in recruiting hundreds of Europeans and North Americans to come and fight in Iraq and Syria is an ample evidence of the success of this narrowcasting tactic.

Chapter 3

Lone Wolves in Cyberspace

The most likely scenario that we have to guard against right now ends up being more of a lone-wolf operation than a large, well-coordinated terrorist attack.

—President Barack Obama, August 16, 2011

The metaphor of the lone wolf terrorist relies on the image of lone wolves in nature. But as this chapter will demonstrate, wolves never hunt alone—in nature or in terrorism. In fact, wolves are one of the most highly social carnivores. In nature, they hunt in packs—groups of animals that are usually related by close blood ties. They live, feed, and travel in these packs. One may not always see the entire group, but their attacks rely on a well-coordinated circling and cornering of the victim. Lone wolf terrorists also

---

This chapter expands on material originally published in "Lone Wolves in Cyberspace" (Weimann 2012a) and "Virtual Packs of Lone Wolves" (Weimann 2014b).

have their pack: a virtual pack. These terrorists are recruited, radicalized, taught, trained, and directed by others. The wave of lone wolf attacks has been propelled by the impact of online platforms, which provide lone wolves with limitless opportunities. Online, an aspiring terrorist can find everything from instructions on building homemade bombs to maps and diagrams of potential targets. In addition, websites, blogs, Facebook pages, and chatrooms all provide easy venues for cultivating extremism in a way that was previously possible only through in-person gatherings.

The Growing Threat

We face threats from those who self-radicalize to violence, the so-called "lone wolf," who did not train at an al Qaeda camp or overseas or become part of an enemy force, but who may be inspired by radical, violent ideology to do harm to Americans. In many respects, this is the terrorist threat to the homeland . . . that I worry about the most; it may be the hardest to detect, involves independent actors living within our midst, with easy access to things that, in the wrong hands, become tools for mass violence.

—Jeh Johnson, Secretary of Homeland Security, February 7, 2014

Lone wolf terrorism is the fastest growing form of terrorism. Before 9/11, the men who went to terrorist camps and jihadi mosques where radical imams preached jihad were seen as constituting the largest terror threat. Since 9/11, a gradual change has occurred. The real threat now comes from the single individual, the "lone wolf," who is living next door, being radicalized on the Internet, and plotting strikes in the dark. A lone wolf is an individual or a small group of individuals who uses traditional terrorist tactics—including the targeting of civilians—to achieve explicitly political or ideological goals, but who acts without membership in, or cooperation with, an official or unofficial terrorist organization, cell, or group. The Unabomber (who carried out attacks from 1978 to 1995), the 1995 Oklahoma City bombing, the 2009 Fort Hood shooting, and the 2011 Norway attacks are examples of this new form of terrorism. Currently, less than 2 percent of terrorism in most countries that keep terrorism statistics can be attributed to lone wolf terrorists—however, the problem is rapidly growing (Spaaij 2010). Acts of lone wolf terrorism have been reported in Australia, Canada, Denmark, France, Germany, Italy, the Netherlands, Norway, Poland, Portugal, Russia, Spain,

Sweden, the United Kingdom, and the United States. The report issued by the Institute for Safety, Security and Crisis Management on Terrorism (COT/ TTSRL, 2007), which uses the RAND Terrorism Knowledge Base, reveals two trends. First, the number of lone wolf attacks has increased in recent decades. And second, lone wolves seem to come from all types of extremist ideological and religious groups.

A recent study on lone wolf terrorism reveals alarming trends. Using a dataset created from RAND, START, and LexisNexis Academic databases, Sara Teich (2013) identified the following trends: (1) An increasing number of countries were targeted by lone wolf terrorists; (2) lone wolf terrorists were responsible for an increasing number of fatalities and injuries; (3) the United States, the most-targeted country, is the target of 63 percent of all global lone wolf attacks (between 1990 and 2013), followed by the United Kingdom, Germany, and other Western countries; (4) compared with other types of terrorists, lone wolves have a higher prevalence and success rate; and (5) military personnel are increasingly the targets of this type of terrorism.

The term "lone wolf" encompasses a broad range of terrorists and motivations. The concept was popularized in the late 1990s by two white supremacists, Tom Metzger and Alex Curtis, as part of an effort to encourage fellow racists to act alone in committing violent crimes for tactical reasons. Other terms that have been used to describe similar or comparable forms of political violence include "leaderless resistance" and "freelance terrorism." One definition of lone wolf terrorism is as follows: ". . . terrorist attacks carried out by persons who (a) operate individually, (b) do not belong to an organized terrorist group or network, and (c) whose *modi operandi* are conceived and directed by the individual without any direct outside command or hierarchy" (Spaaij 2010, 854–55). This definition describes Islamist lone wolves such as Nidal Hasan (the 2009 Fort Hood killer) and Abdulhakim Mujahid Muhammad (who opened fire on a US military recruiting office in Little Rock, Arkansas, in 2009), but also describes anti-Semitic devotees like Buford Furrow (who attacked a Jewish community center in Los Angeles in 1999), and Eric Rudolph, also known as the Olympic Park Bomber (who perpetrated a series of anti-abortion and anti-gay bombings across the southern United States between 1996 and 1998 that killed two people and injured at least 150 others).

In December 2013, the al-Qaeda media outlet As-Sahab released a two-part video calling for lone wolf attacks in the West and highlighting individuals such as 2009 Fort Hood shooter Nidal Hasan as examples to follow.

The video, an updated version of the 2011 propaganda movie "La Tuka-lifu Ila Nafsak" (Commit No One But Yourself), referenced the April 2013 Boston bombings and the May 2013 Woolwich attack (in which an off-duty British Army soldier was murdered on the street near his barracks) as additional models for Muslims to follow. According to the text, "The Boston attacks brought a distant war back home to the heart of America. . . . These attacks left a message for the American public that the $60 billion budget of the Directorate of National Intelligence, besides the trillions of dollars of taxpayers' money spent on the imperial wars abroad as well as on homeland security, failed to protect them from a simple attack executed by two men who managed to breach America's defenses, despite being on the FBI's watch list since 2001. With the Boston attacks, the notion of an impregnable America in the post-9/11 era also evaporated into thin air" (MEMRI Jihad & Terrorism Threat Monitor 2011a). The video also included an explanation for the use of lone wolf attacks:

> The causes which necessitate adopting this form of warfare that depends on an individual or a small group of individuals for its execution are diverse. These include:
> - The virtue of this type of jihad in our religion.
> - Following our virtuous predecessors.
> - Expansion of the theatre of war.
> - The onslaught of the enemy against the Ummah from all directions.
> - Dispersed interests of the enemy, whether in the enemy's own land or Muslim lands.
> - Ease of targeting such enemy interests and its relatively huge impact. (MEMRI Jihad & Terrorism Threat Monitor 2011a)

The lone wolves are challenging the police and intelligence community, as they are extremely difficult to detect and to defend against. Compared to group or network terrorism, lone wolves have a critical advantage: they easily avoid identification and detection before and after their attacks, since most do not reveal their inclinations, visions, and plans. However, they are not completely divorced from contact with others. They connect, communicate, and share information, know-how, and guidance—all online—on the "dark web."[1]

---

[1] The term "dark web" (also referred to as "deep web") stands for hard-to-find websites and secretive networks that sometimes span across the Internet and can be used for criminal activities or terrorism.

The Virtual Pack

The Internet has proven vital to every type of lone wolf. According to Marc Sageman (2008), most lone wolves are part of online forums, especially those who go on to actually carry out terrorist attacks. This makes the Internet an incredibly important source for finding potential lone wolf terrorists. Colleen LaRose, also known as "Jihad Jane," used MySpace, YouTube, and email to contact other extremists and express her desire to become a martyr for the Islamic cause. In August 2009, she traveled to Europe to take part in an assassination plot against Lars Vilks, a Swedish illustrator who had angered Muslims throughout the world with his derogatory caricature of the Prophet Muhammad. In 2010, Roshonara Choudhry attempted to assassinate Stephen Timms, a British member of Parliament who had supported the 2003 Iraq war. Choudhry had showed no signs of radicalization prior to the attack and gave no indication to friends, family members, or acquaintances that she sympathized in any way with Islamic extremism. But for months she had been secretly downloading sermons by Anwar al-Awlaki—more than 100 in all. She never met, emailed, or talked to al-Awlaki, but was motivated solely by his online calls for violence against the West (Collins 2010).

In addition to giving the possibility of becoming part of a community, the Internet is a platform where lone wolves can express their views. The 2010 Stockholm suicide bomber Taimour Abdulwahab al-Abdaly, for example, was active on the Internet. He had a YouTube account, a Facebook account, and even searched for a second wife on Islamic Web pages. Anders Behring Breivik, the far-right perpetrator of the 2011 Norway attacks that killed 77 people, used several different social networking sites such as Facebook and Twitter, and posted his manifesto "2083—A European Declaration of Independence" on the Internet before committing two terror attacks in downtown Oslo and on the nearby island of Utøya.

What is the role of the Internet as an incubator or accelerator of the lone wolf phenomenon? Raffaelo Pantucci's (2011, 14) conclusion in this regard is instructive:

> The internet is clearly the running theme between most of the plots included in this dataset [on lone wolves] and it appears to be a very effective tool: it provides a locus in which they can obtain radicalizing material, training manuals and videos. It provides them with direct access to a community of like-minded individuals around the world with whom they can connect and in some cases can provide them with further instigation

and direction to carry out activities. Many of the individuals in the dataset demonstrate some level of social alienation—within this context, the community provided by the internet can act as a replacement social environment that they are unable to locate in the real world around them.

A report by the Dutch General Intelligence and Security Service (AIVD 2012, 20–21) came to a similar conclusion:

> The AIVD is aware of the fact that lone wolves often plot and carry out a (violent) act on their own, but has found that they rarely radicalize in complete isolation. . . . The AIVD argues that radicalization is a social phenomenon. This also applies to most lone wolves. In the aftermath of such events, it is often discovered that lone wolves hardly had any contact with like-minded individuals in real life, but did maintain active contact with people on the Internet. In retrospect, it is then concluded that these contacts, as well as the consumption of jihadist propaganda and the online discourse, have contributed to their radicalization and (may also) have inspired them to commit such a (violent) act.

A typical example may illustrate the online practice. In 2011, Jose Pimentel was arrested for planning attacks with homemade pipe bombs against police vehicles and postal facilities in New York and New Jersey, as well as against US troops returning from Iraq and Afghanistan. He was charged with providing support to terrorism, conspiracy, and weapons offenses. In February 2014, Pimentel admitted to plotting a terrorist attack using a pipe bomb (Investigative Project on Terrorism 2014). Pimentel, a native of the Dominican Republic who had come to the United States at the age of 8, was an unemployed, 27-year-old Muslim convert and al-Qaeda sympathizer who lived in Manhattan. He was not part of any known al-Qaeda organization, but the inspiration for his planned bombing attacks came from reading instructions in the online *Inspire* magazine produced by al-Qaeda and the American-born cleric Anwar al-Awlaki. According to New York authorities, Pimentel was not just radicalized by the al-Qaeda magazine, but he also found in its pages the instructions he used to build the pipe bomb, thanks to the magazine's notorious English-language article, "How to Make a Bomb in the Kitchen of Your Mom." Moreover, reading *Inspire* magazine was not Pimentel's main online activity. He maintained a massive website on Blogger and a YouTube channel featuring hundreds of radical works. An analysis of Pimentel's online footprints reveals that he was "directly linked online to known extremists through whom he is connected to some hundreds of like-minded individuals" (Internet Haganah 2011). Pimentel's website, TrueIslam1.com, hosted an impressive archive of jihadist texts, with audio and video organized by means of the online publishing tool Blogger. The website connects to Pimentel's YouTube channel, which was similarly thorough; it had collected more than 600 videos related to radical and violent interpretations of Islam, 60 of which he had uploaded himself. This channel had more than 1,500 subscribers.

Al-Qaeda Calling the Lone Wolves

One of the most difficult challenges faced by al-Qaeda today is the ongoing loss of a large part of its first-, second-, and even third-generation leadership, some of whom have been assassinated or arrested, while others have dissociated themselves from the organization and its terrorist methods. As observed in a recent Europol report, "As a consequence of sustained military pressure, al-Qaeda core have publicly discouraged sympathizers from travelling to conflict zones in order to join them. It has instead promoted the idea of individually planned and executed attacks in Western countries without the active assistance of any larger organization" (Europol 2012, 17). Partly out of necessity, al-Qaeda has now thrown its weight fully behind "lone" terrorism. As early as 2003, an article was published on the extremist internet forum Sada al Jihad (Echoes of Jihad), in which Osama bin Laden sympathizers were encouraged to take action without waiting for instructions. In 2006, a text authored by al-Qaeda member Abu Jihad al-Masri, "How to fight alone," circulated widely in jihadist fora. Another prominent Salafi writer, Abu Musab al-Suri, also advocated that acts of terrorism be carried out by small, autonomous cells or individuals. He outlined a strategy for global conflict that took the form of resistance by small cells or individuals and kept organizational links to an absolute minimum.

In March 2010, al-Qaeda's As-Sahab released an English-language video titled "A Call to Arms," which featured an American-born spokesperson, Adam Gadahn. The video, directed at jihadists in Israel, the United Kingdom, and the United States, highlights the 2009 Fort Hood shooter, Nidal Hasan, whom Gadahn describes in glowing terms ("a pioneer, a trailblazer and a role model who has opened a door, lit a path, and shown the way forward for every Muslim who finds himself among the unbelievers"). Hasan is held up as an exemplary figure for his loyalty to Islam and Muslims in defiance of his unbeliever commanders and for having struck at a sensitive target in the heart of America. Gadahn then uses the example of Hasan to call on other Muslims in the "Crusader West," especially in Israel, the United Kingdom, and the United States, to undertake lone wolf attacks. He advises his listeners to focus in particular on targets that will do serious economic damage to these countries, and points to the 9/11 attacks to show that such attacks need not employ conventional firearms. In an early June 2011 English-language video message, headlined, "Do Not Rely on Others, Take the Task upon Yourself" (also known as "Commit No One But Yourself," as mentioned above), Gadahn emphasizes lone wolf operations even more clearly. He suggests possible measures: "Let's take America as an example. America is absolutely awash with easily obtainable firearms. You can go down to a gun show at the local convention center and come away with a fully automatic assault rifle, without a background check, and most likely without having to show an identification card. So what are you waiting for?" (quoted in Cole 2011).

Al-Qaeda in the Arabian Peninsula (AQAP), an al-Qaeda affiliate, has been especially vocal in encouraging lone acts of terrorism. Its online English-language *Inspire* magazine promotes "open-source jihad," a new tactic that emerged as the al-Qaeda leadership steadily vanished in the decade following 9/11. The organization splintered first into "franchises" by country or region, then further into lone operators. *Inspire* became an important tool for recruiting, informing, and motivating these lone jihadists. Each edition of the magazine, in fact, has a special section called "Open Source Jihad," which is intended to equip aspiring jihadist attackers with the tools they need to conduct attacks without traveling to jihadist training camps. It is also dedicated to helping terrorist sympathizers in the West carry out attacks by including, among other things—as was seen above in the case of Jose Pimentel—bomb-making recipes. *Inspire* has featured articles attributed to three prominent violent jihadist propagandists with strong American ties: Gadahn, the radical American-born imam Anwar al-Awlaki, and the Saudi-born American citizen Samir Khan. The latter two were killed in a US air strike in Yemen in September 2011. The article "How to Make a Bomb in the Kitchen of Your Mom," mentioned earlier, was downloaded by individuals who plotted terrorist attacks in both the United Kingdom and the United States—including Naser Jason Abdo, a Muslim US soldier who allegedly plotted to attack the Fort Hood military base in 2011, and Jose Pimentel, who had started making a pipe bomb based on the recipe when he was arrested.

Inspire's articles clearly promote individual jihad; thus, the fall 2010 edition editorialized: "Spontaneous operations performed by individuals and cells here and there over the whole world, without connections between them, have put the local and international intelligence apparatus in a state of confusion, as arresting the members of aborted cells does not influence the operational activities of others who are not connected with them." The ideas and methods for terror attacks are meant for anyone, including those without direct ties to al-Qaeda or its affiliates. For example, the summer 2010 issue advised making a pipe bomb using everyday materials; the fall 2010 issue encouraged using one's car to "mow down" people in crowded places, and the winter 2010 issue discussed how to blow up buildings.

Since its foundation, *Inspire* magazine has also advocated the concept that jihadists living in the West should conduct attacks there, rather than traveling to places like Pakistan, Somalia, or Yemen, since such travel might bring them to the attention of the authorities. Indeed, *Inspire* views attacking in the West as "striking at the heart of the unbelievers." An October 2010 issue article entitled "Tips for Our Brothers in the United States of America" contained this recommendation: "We strongly encourage our brothers to fight jihad on US soil. . . . A random shooting rampage at a crowded restaurant in Washington D.C. at lunch hour, for example, might end up knocking out a few government employees [and it] would attract additional media attention." Gaining world attention has always been important for al-Qaeda.

On May 2, 2012, the eighth and ninth issues of *Inspire* were made available through al-Qaeda's jihadist forums. Both editions reinforced promotion of lone wolf attacks, but each issue presented different arguments and directions. The eighth issue, with the cover headline "Targeting Dar al-Harb Populations," advocated the lone wolf model for non-Muslim lands in the same way that previous editions did. It detailed plans for new attack methods in the "Open Source Jihad" section, and presented the culmination of Anwar al-Awlaki's justification for killing American civilians. This issue also explained to potential lone wolves how to use small handguns and how to build remote-controlled detonators for explosives. The contents of the eighth issue clearly relied on articles from previous issues. The ninth issue, entitled "Winning on the Ground," included instructions for individuals wishing to carry out lone wolf jihad attacks. An article entitled "The Convoy of Martyrs: Rise Up and Board with Us" declared,

The objective of this workshop is to communicate with those seek[ing] martyrdom operations, or those who want to execute a slaughter to the enemies of Islam, [or] those who have no means of contacting their *muja-hideen* brothers. Whatever the reason, the aim is to activate them in the midst of the enemy, weather [sic] the enemy is the Jews, the Christians or the apostates. It is becoming obvious to many that the concept of individual jihad, which [has] begun to appear recently, has been called for by the leaders of jihad.

The article instructed candidates to send basic information about themselves to AQAP's "military committee," which will then help them plan and execute the attack; it will also take responsibility for the incident and provide media coverage. The article included a list of conditions that candidates must meet, a list of possible targets, and directions for contacting the committee securely by using encrypted emails. The ninth issue encouraged individuals to act alone in gathering information, preparing, and ultimately executing an attack. Although the focus was still on individual jihad, the article did argue that there must be an operational leadership, which AQAP will provide. This issue also placed strong emphasis on lone wolf operations to limit the opportunity for law enforcement interdiction, and advised small-group plots only if the individual completely trusts his associates. The article "Convoy of Martyrs" described the qualifications of the desired lone wolf: you must be a Muslim, possess "maturity," and be skilled in "listening and obeying." The terrorist group provided a public encryption key and a handful of Gmail, Yahoo!, and Hotmail accounts where potential lone wolves could send ideas about whom or what they would like to attack. Once the idea is approved, the lone wolves would be sent to act, unencumbered by any traditional terrorist cell: "The only connection that mujahid has," the text stated, "is with the group leadership. In this case it will be our military committee."

There is convincing evidence of the impact of *Inspire* magazine among lone wolves. A growing number of lone wolves were found to be linked with the magazine. US examples include Jose Pimentel and Naser Jason Abdo, and British examples include university dropout Hakan Ertarkan, who was arrested in London on April 12, 2011, and found to be in possession of a CD issue of *Inspire*. Similarly, German nationals Christian Emde and Robert Baum were arrested on July 15, 2011, when they tried to enter Britain in possession of electronic copies of *Inspire*. Also in Britain, Zahid Iqbal, Mohammed Sharfaraz Ahmed, Umar Arshad, and Syed Farhan Hussain were arrested on April 24, 2012, and accused of working to recruit others (drawing their inspiration from *Inspire*), and of possessing copies of the magazine.

The Boston Marathon bombers of April 2013 were also linked to *Inspire*. In the weeks following the attacks, Dzhokhar Tsarnaev told investigators that he and his brother learned to make pressure-cooker bombs from the magazine (Horwitz 2013). They seem to have followed *Inspire*'s tips, using gunpowder emptied from fireworks, shrapnel glued inside a pressure cooker, and a commercial remote control as detonator. "The pressurized cooker should be placed in crowded areas and left to blow up," the manual says. "More than one of these could be planted to explode at the same time." The younger Tsarnaev also said that he felt inspired by the online sermons of al-Qaeda propagandist Anwar al-Awlaki, who edited the magazine (Finn 2012). Although the instructions for making the pressure-cooker bombs were copied from *Inspire*, the target and style of the attack also mirrored instructions released by one of al-Qaeda's leading strategic thinkers and trainers, Abu Musab al-Suri. In selecting the Boston Marathon for their attack, the actions of the brothers continued to reflect Suri's instructions for "solo jihad," published in *Inspire*. Al-Qaeda's online propaganda machinery made much of this connection: the eleventh issue of the *Inspire* magazine, issued in June 2013, devoted almost all of its 40-odd pages to glorifying what it calls the "BBB"—the "Blessed Boston bombings." In an article on the links between *Inspire* and the Boston bombings, the magazine's editor Yahya Ibrahim noted: "Yes, the brothers have been inspired by *Inspire*. This is not only because *Inspire* offers bomb recipes, but also because of the contents of the magazine as a whole."

Rather than requiring its readers to seek out individual issues, the producers of *Inspire* magazine have even made it easier to find its complete contents. In March 2013, a month before the Boston attack, the al-Qaeda branch in Yemen posted the "Lone *Mujahid* Pocketbook," an online compilation of all of *Inspire*'s do-it-yourself articles in English combining high-quality graphics and teen-friendly style. The 64-page manual asks "R U dreamin' of wagin' *jihadi* attacks against *kuffar* (non-believers)?" "Have u been lookin' 4 a way to join the *mujahideen* in frontlines, but you haven't found any?" "Well, there's no need to travel abroad, because the frontline has come to you" (Shane 2013).

The Online Recruitment Process

The recruitment of lone wolf terrorists has several distinctive characteristics. First, it is a multistep process, which requires a gradual transition and numerous phases. Second, unlike conventional or traditional recruitment, the process relies on online platforms. The study of lone wolf recruitment reveals a few common patterns and stages.

Applying the RAND Corporation's model of selection and recruitment, the first step is "the Net": A target population may be engaged equitably by being exposed to an online message, video, taped lecture, or the like (Gerwehr and Daly 2006). Some members will respond positively, others negatively; but in general, the whole population is viewed as primed for recruitment. More specifically, the target audience is viewed as homogeneous enough and receptive enough to be approached with a single undifferentiated pitch. For this "netting" stage, all online platforms may be used—from Facebook pages to personal mail, from YouTube video clips to Twitter or official websites. (Chapter 10 presents the seductive narratives used by terrorist groups to lure and persuade individuals who are socially frustrated, alienated, hopeless, or angry.) The techniques of seductive narratives rely on well-known content elements that activate processes of identification and transformation. In March 2014, AQAP released the twelfth issue of *Inspire*, featuring articles promoting arson and car bombs to strike the enemy. In its effort to convince Western Muslims to carry out lone wolf attacks, AQAP highlighted the impact that these attacks have on the United States and provided would-be recruits with practical information and advice, including detailed, illustrated instructions for building a car bomb, and a list of recommended targets. Throughout the magazine, AQAP also provided would-be attackers with religious, ideological, and moral justification for such attacks. Citing the deceased Anwar al-Awlaki, one of AQAP's main recruiters of Western operatives, the organization called for and justified attacks on civilian targets to produce the greatest possible impact and damage. The magazine quoted an email exchange between al-Awlaki and one of his followers in the West, in which the former wrote: "If you ask me as a tactic, is targeting the civilian population of the West a good thing to do? I would say yes, because it is much more potent and powerful . . . when you hit the civilian you hit them where it hurts most and that is what our tactics are about." Al-Awlaki even sanctioned the killing of women and children: "When men, women and children are mixed and integrated such as in a city or village there is no doubt that it is allowed to target them while carrying the intentions of not specifically targeting the women and children. Therefore, an attack on a population center such as a US, British, French or German city with a bomb or a firearm attack is definitely allowed."

The second stage is the "funnel." As the term implies, potential recruits start at one end of the process and are transformed, after some culling along the way, into dedicated members when they emerge at the other end. Here, the recruiter may use an incremental, or phased, approach when he or she believes a target individual is ripe for recruitment yet requires a significant transformation in identity and motivation. This stage capitalizes on a wealth of well-studied techniques in cognitive, social, and clinical psychology. It involves online exchanges and exposure to religious, political, or ideological material. This stage relies on social bonding (albeit a virtual one), based on the target's alienation, social frustration, solitude, and personal pessimism.

The next stage is the "infection." Selected target members who are dissatisfied with their social status or have a grudge against their political or religious system are directed to self-radicalization. The self-radicalization of lone wolves relies only on online sources and involves gradual advancement in the level of commitment and extremism. It often relies on the "seed crystal" practice: the process may be compared to lowering the temperature of a glass until the water inside it cools and ice crystals begin to form as the seeds of a complete freeze. In "seed crystal" recruitment, critical variables include the type of various forces being used to "chill the glass," and the rigidity of the "freeze." These forces may include an advanced radicalization by continuous exposure to online radical material and by virtual online guidance. For al-Qaeda, the seed crystal approach is the most successful in diasporas or populations where open recruiting is difficult or impossible.

The final stage, the "activation," involves the release of the lone wolf to carry out the terrorist action. This stage includes practical instructions, such as online manuals on the use of explosives, weapons, poisons, and chemicals; directions regarding the selection of the target and the location and timing of the attack; and the final send-off.

Chapter 4

The E-Marketing of Terror

What Is E-Marketing?

E-marketing refers to the use of the Internet and digital media capabilities to help sell products and services. It is based on online interactive technologies used to create and mediate dialogue between firms or companies and identified customers (Coviello et al. 2003). This method is also known as i-marketing, Internet marketing, web marketing, online marketing, or online advertising. It includes email marketing, search engine marketing, social media marketing, many types of display advertising (including web banner advertising), and mobile advertising. E-marketing is growing rapidly. In 2011, US Internet advertising revenues surpassed those of cable television and nearly exceeded those of broadcast television, and in 2012, US Internet advertising revenues totaled $36.57 billion, a 15.2 percent increase over the

This chapter expands on material originally published in "The E-Marketing Strategy of Hamas" (Mozes and Weimann 2010).

$31.74 billion in revenues in 2011 (Interactive Advertising Bureau 2013). As with conventional marketing, e-marketing is a strategy that helps to deliver the right messages, products, and services to the right audience. It consists of all activities and processes involved in finding, attracting, winning, and retaining customers. What makes e-marketing different is its wider scope and options compared to conventional marketing methods, and its ability to be broken down into specialized areas such as web, email, and social media marketing:

- Web marketing includes e-commerce websites, affiliate marketing websites, promotional or informative websites, online advertising on search engines, and organic search engine results via search engine optimization.
- Email marketing involves advertising and promotional marketing efforts through email messages to current and prospective customers.
- Social media marketing involves advertising and marketing (including viral marketing) efforts through social networking sites such as Facebook, YouTube, Twitter, and Digg.

Paul Russell Smith and Dave Chaffey (2001) noted that Internet technology can be used to support marketing aims in the following ways:

- Identifying: The Internet can be used for marketing research to determine consumers' needs and wants.
- Anticipating: The Internet provides an additional channel by which consumers can access information and make purchases. Understanding this demand is key to governing resource allocation in e-marketing.
- Satisfying: A key success factor in e-marketing is achieving customer satisfaction through the electronic channel, raising such issues as "Is the site easy to use?", "Does it perform adequately?", and "Does it fulfill the consumers' needs?"

E-marketing enjoys several advantages for commercial enterprises (as well as for terrorist groups). The following benefits may explain the rising expenditures on e-marketing:

- Wider prospect reach: The Internet has become part of everyone's life. Whatever products are offered, there is an existing market online. E-marketing makes it easier to find new markets and potentially compete worldwide with only a small investment.

- Cost-effective approach: A properly planned and effectively targeted e-marketing campaign can reach target customers at a much lower cost than traditional marketing methods.
- 24/7 marketing: With online presence, there is no need for physical presence or physical premises while the operation runs continuously.
- Personalized one-on-one marketing: E-marketing creates interactive, direct channels of communication that promote personal exchanges.
- Multimedia interactivity: E-marketing combines text, graphics, music, and videos. Through two-way communications and interactive multimedia platforms, one can engage with wide audiences and give them greater involvement and control.
- Increased ability to track results: E-marketing makes it easier to measure a campaign's effectiveness. It allows the users to obtain detailed information about customers' responses to the campaign, such as through the use of methods such as pay per click or pay per action.

Numerous studies on e-marketing have yielded practical guidelines and directions for effective online marketing. The importance of developing an effective e-marketing strategy is indicated by Michael Porter (2001), who argued that the key question is not *whether* to deploy Internet technology—because companies have no choice if they want to stay competitive—but *how* to deploy it. There is no evidence to suggest that the approach to developing and implementing strategy should be significantly different for e-marketing than for traditional forms. Established frameworks for corporate strategy development or strategic marketing planning should still be followed. These frameworks provided a logical sequence to ensure that the key activities of strategy development are included. However, with e-marketing there is an even greater need for a highly responsive strategy process model that can react quickly to events in the marketplace. The preferred approach is an emergent e-marketing strategy process that involves continuous improvements and revisions (Chaffey 2004).

One of the leading textbooks in this developing area is *Internet Marketing: Strategy, Implementation and Practice* (Chaffey et al. 2000). When considering or evaluating e-marketing strategy, the authors suggest a model based on the following eight decisions:

- Decision 1: Who are the potential audiences?
- Decision 2: Positioning and differentiation
- Decision 3: Resourcing
- Decision 4: Migrating the company's brand to the Internet

- Decision 5: Outsourcing and strategy partnerships
- Decision 6: Organizational support
- Decision 7: Building a budget and resource allotment
- Decision 8: Channel structure modifications. (Chaffey et al. 2000, 174–203)

Terrorists and E-Marketing

Is the strategy of e-marketing applicable to the analysis of terrorist online presence? Can we use the models developed for commercial websites to analyze terrorist websites? We attempted to apply the notion of e-marketing and its practices to terrorist websites (Mozes and Weimann 2010). The analysis focused on Hamas's online presence, applying the eight-decision-point model offered by Dave Chaffey and his colleagues (2000) in order to understand the set of considerations and decisions that guided the design of Hamas websites and to evaluate their fit within an e-marketing model. It should be noted that the eight-point model was conceptualized to direct the strategy of constructing websites and to help decision makers make the right choices at each stage of the process. In our study, the process is in fact reversed. The websites already exist on the Internet, and the model is used to assess the strategic decisions behind their content and format.

The following analysis is based on the study of Hamas websites, initiated and maintained by the Palestinian Information Center (PIC). Hamas relies heavily on the Internet to spread its messages. Hamas operates numerous websites, including official websites; forums; chatrooms; online bulletins; video clip sites; and special websites for children, youth, and women. PIC was established in December 1997 and states on its English-language pages that it is "an independent Palestinian organization." The Arabic version of its website, however, describes itself as "the mouthpiece of resistance, Hamas and Jihad." The PIC website, which provides links to various Hamas-affiliated websites, features the latest issues of *Qassamyoon*, the magazine of the Izz ad-Din al-Qassam Brigades, the military arm of Hamas. The magazine glorifies terrorism and includes military instructions on how to engage with "the enemy."

Decision 1: Who Are the Potential Audiences?

In Chaffey and colleagues' model (2000), the first key decision point concerns the target audience. This decision involves evaluating and selecting appropriate segments and developing appropriate offers. In an Internet context, organizations typically target those customer groupings with the highest propensity to access, choose, or buy online. The e-marketing model suggests three different parameters to examine the target audience and to reach a decision about the type of services that each target audience will receive. First, Chaffey and his colleagues differentiate between the types of potential audiences: consumers, personnel, and third parties. They advise companies that a successful website should provide the three types of users with suitable functions in the website. In the next step, they suggest focusing on different types of customers: the most profitable, the largest, those hard to reach using other kinds of media, those that do not have brand loyalty, and strategic decision makers. The last reference point they emphasize is the topic of "localization," or addressing the needs of different audiences in different countries (Chaffey et al. 2000, 317).

Although the official Hamas website run by the PIC is designed to target all three potential audiences, it addresses mainly the first type of users: consumers. The first page of the PIC site offers services to users in eight different languages. A comparison of the site pages in the different languages reveals differences across the websites, even though visually they all appear similar.

- *The English site:* There is no connection between the website and the Hamas organization or its factions. The links shown on the site send the user to Western news websites or to Palestinian human rights websites. Furthermore, a search for the word "Hamas" in the English website will reveal only news headlines where Hamas spokesmen express their opinions about current affairs. The entire website does not include any information about the organization, its activities, or its casualties. The degree of visual extremism is very low.
- *The Russian site:* It is hard to find a clear connection to the Hamas organization, except for one picture where the Dome of the Rock is clearly shown and against the background, in the sky, is a barely visible photo of the founder of the organization, Sheik Ahmed Yassin. The amount of visual extremism is very low and is limited to pictures and cartoons with anti-Semitic undertones.
- *The Urdu site (Pakistan):* The site does not present a visual connection to Hamas; yet the degree of visual extremism is high. For example, a cartoon on the site shows a dove of peace with its throat slit and blood spilling from it. The site has a banner that calls for a boycott of Israeli products and features a young girl crying.

- *The Turkish site:* The site's affinity to the Hamas organization is high and is expressed through a short Flash animation where different sentences appear on a clear background picture of Sheikh Yassin's face. The degree of visual extremism is slightly higher than the English, Russian, or Urdu sites—one banner shows a crying girl, and another commemorates 60 years since the Palestinian Nakba (meaning "day of the catastrophe," the annual day of remembrance for the Palestinian people of the "anniversary of the creation of Israel" in May 1948).
- *The Indonesian site:* The site shows a clear connection to the Hamas organization using a propaganda poster for "Iz A Din Al Kassam," a variant name for Hamas' operational military division, the Izz al-Din al-Qassam Brigades. In addition, clicking on the photo will lead users to the operational division website. The degree of visual extremism is moderate.
- *The French site:* This site presents more extreme visual content and there is a clear connection to Hamas. The affinity of the site to the Hamas organization is clear from the fact that the Hamas logo is visible on the homepage. Pictures of Sheikh Yassin and Abed al-Aziz al-Rantisi, labeled as "martyrs," are presented. The high visual extremism of the site is expressed through pictures of bloody bodies of young men, corpses, and a picture of a crying baby with her face covered in blood.
- *The Persian site:* In this site, the connection to Hamas is accentuated. The Hamas logo is shown, and the operational division logo is shown as well, with a link to another Hamas site—"Palestine in voices and pictures." The Persian homepage does not show extreme photos, except for anti-Semitic, anti-Western cartoons. It should be noted that Iran is a state-sponsor of Hamas.
- *The Arabic site:* This is in fact the main site among the numerous PIC websites. In this site, the connection to the Hamas organization is the clearest. The Hamas logo and the operational division logo are both shown clearly. The site links to the *Al-Fateh* children's online magazine and to Hamas's numerous forums. The degree of visual extremism is not especially high, and except for some blatant cartoons along the lines of those mentioned above, there are no photos that evoke strong emotions.

Thus, the comparison reveals significant variance in the websites' contents, in accordance with the Chaffey model. The various languages used indicate different target audiences and a variance in style, text, appeals, and visuals, demonstrating the communicators' awareness of their different audiences. Moreover, based on one marketing model, one can further divide the different audiences targeted by the PIC websites:

1. Audiences that are Hamas-oriented: People who openly support Hamas and who look for information about the organization as a social organization, a Palestinian political movement, or an active militant unit;
2. Audiences that are generally interested in the Palestinian issue: This audience shows interest in Palestinian history or politics and seeks information about Palestinians, their conflict with Israel, and their background;
3. Audiences that are anti-Western: People who are opposed to the West; these include anti-Americans and very often also include anti-Semites; and
4. Audiences that are interested in human rights: This audience shows an interest in human rights violations.

The different PIC sites direct specific messages to the above groups. All the sites have some common elements: They feature contents that present Palestinian issues, such as articles about inter-Palestinian political issues, and articles about the Palestinian struggle against Israel and its supporters. However, the sites differ in terms of the degree to which they pander to an anti-West audience or to a human rights–focused constituency. An anti-West audience can be targeted with human rights messages, but an appeal to the human rights audience cannot involve anti-West messages. This stems from the understanding that Western human rights activists will be deterred from the message of websites that have a strong anti-West orientation.

Additional comparison across target audiences involved the analysis of the same website in two languages: English and Arabic. Such comparison reveals interesting differences. All the PIC sites offer "A Look at History" where a visitor can enter a date and receive a detailed list of events relating to Palestinian issues over the years. We compared the information provided when specific dates were entered. Thus, when the date "March 30" was entered, the English version presented three events that occurred on that day: In 1989, "Israel bars UNRWA from running an informal educational program in the West Bank for kids that were left idle after the military closure of schools"; in 1987, "On a television interview the Israeli minister of war, Rabin, exposes the fact that 1000–4400 Palestinians are imprisoned in Israeli jails"; and in 1976, "Israeli police forces and soldiers kill 6 Palestinians living in the occupied territories since 1948 during a demonstration against the occupation of the lands by Israel." However, the Arabic version presents five different events: In 2003, "A sacred suicide attack in Netanya, Israel" (the name and hometown of the attacker are mentioned); in 2002, "The Zionist occupation acts in Ramallah, Bethlehem, and other cities"; in 2002, "A sacred suicide attack in Tel Aviv" (the name and hometown of the attacker are again mentioned); in 2001, "the Zionist occupation forces injured demonstrators on 'Land Day' in Nablus"; and in 1976, "Proclamation of Land Day on the lands of occupied Palestine."[1]

Additional comparison between the English and Arabic websites examined the articles posted on these sites. This comparison revealed that both sites presented articles dealing with inter-Palestinian political issues, Israeli attacks in the Palestinian territories, and the vulnerability of Palestinian president Mahmoud Abbas in the Palestinian territories. However, only the English site posted articles on "Zionist Terrorism," "Palestinian Suffering," and "The Injustices Made by the Government of Mahmud Abbas." The Arabic site posted articles missing in the English version, such as those on the resistance in Lebanon and on Israel's weak position in the world.

These and other comparisons between the English and Arabic versions lead to the conclusion that the English site appeals more to human rights audiences and to global audiences interested in Palestinian issues. The numerous articles about injured civilians and the presentation of inter-Palestinian issues that criticize the Fatah government support this conclusion. In the English site, Hamas' anti-West messages are completely absent. The main message conveyed to this audience is "exposing the truth" about what is going on in Palestine, while emphasizing issues that are in contrast to accepted Western notions of "the good side" and "the bad side," all the while concealing any affiliation with the Hamas movement. In contrast, the Arabic version clearly targets Palestinian audiences: it focuses on social and economic issues of Palestinians and posts many references to Hamas social and welfare activities. The differentiation according to target audiences is revealed also in the other versions (languages) of the PIC websites. The Russian site, for example, highlights anti-West messages and also caters to anti-Semitism. Similarly, the Turkish site criticizes the involvement of

---

[1] On March 30, 1976, Palestinian mass protests over the Israeli government's decision to expropriate Arab land in the West Bank escalated into violent confrontations in which six Palestinians were killed. The annual Palestinian commemorations of "Land Day" have different meanings for Israelis and Palestinians (see Wolfsfeld, Avraham, and Aburaiya 2001), as reflected in the different versions of events on the PIC sites.

Israel and the United States in Turkish politics, expressed in the empower-ment of the Turkish military—the strongest secular power in the country. In a more extreme way, the French site clearly targets a Muslim audience with a strong anti-West orientation. The site emphasizes graphic pictures that accompany written accounts of Israeli military actions, underlines the French president's opinions against Muslims and Palestinians, and empha-sizes the Israeli-French connection through French Jews.

Decision 2: Positioning and Differentiation

Chaffey and colleagues (2000, 180–85) present three levels of positioning information that can be promoted on a website: "brochureware,"[2] "interac-tion," and "representation" (Web self-service). When examining the PIC websites from this dimension, it is possible to see that most of the sites put an emphasis on information presentation according to the brochureware style. The information shown on the websites attempts to be journalistic in nature and to distance itself from propaganda material, in accordance with the differential targeting as explained in the first decision point. More-over, the PIC websites follow the interactive mode: all of the websites have options to register and receive information from the site by email or through its RSS (Really Simple Syndication) feed, a service format for distributing and gathering content from sources across the Web, including newspapers, magazines, and blogs. The contact created by registering and exchanging messages follows the interactive dimension as suggested by the model. The Arabic PIC site has links to Hamas forums and chatrooms, adding one more dimension of interactivity. Hamas supporters use numerous online forums, some of which are operated or monitored by Hamas operatives. Gabriel Weimann (2007a, 2008d) describes the process of online radicalization, especially of youth, by Internet-savvy terrorists. The creation of virtual com-munities, the social bonding online, and the radicalization process—are all instruments of ideological recruitment. These, in fact, are the terrorist ver-sion of the last positioning measure: representation.

Decision 3: Resourcing

Chaffey and colleagues' model (2000) emphasizes an organization's need to decide upon the connection between the Internet presence and the other

---

[2] Brochureware refers to websites or pages that are produced by taking an organization's printed brochure and translating it directly to the Web.

marketing divisions that it operates. Nirmalya Kumar (1999) suggests that a company should decide whether the Internet will primarily complement the company's other channels or replace other channels. Clearly, if it is believed that the Internet will primarily replace other channels, then it is important to invest in the resources, promotion, and infrastructure to achieve this goal. This is a key decision, as the company is essentially deciding whether the Internet is "just another communications and/or sales channel," or whether it will fundamentally change the way that the company interacts with its customers and channel partners (Kumar 1999, 7).

Hamas presents and markets its ideas to the Palestinian audience through a rich variety of platforms, including through sermons in mosques; books, manifests, and brochures; and a variety of electronic media, including radio and television. It is clear that Hamas attempts to integrate all these conventional, traditional, and modern media into one multimedia platform. Most of the Hamas websites are linked to other media outlets operated by the organization. This is evident in the Arabic version, where links to other sites, references to other media, and posting of other Hamas publications are frequently highlighted. In contrast, in the English version, where the affiliation with Hamas is downplayed, such "networking" is absent.

Decision 4: Migrating the Company's Brand to the Internet

Chaffey and colleagues' model presents four options to migrate the brand to the Internet: (1) migrate the traditional brand online, (2) extend the traditional brand, (3) partner with an existing digital brand, or (4) create a new digital brand (Chaffey et al. 2000, 227–29). The Hamas organization, in expanding its activities to the Internet, has chosen two of these options. Regarding the existing media platforms of the organization, its associations and its operational division, it chose the first option—the brands of the organization were shifted to the Internet as they were, including the website address as the chosen brand name. However, while this is true for most of the Hamas websites, the PIC websites are an exception: here, Hamas clearly preferred the fourth option. A new brand was created and, as we have seen, the organization went further and even concealed the connection between the new brand and the traditional brand in various language sites, except for those in Arabic. For instance, since 2012, the PIC has a free application on iTunes that provides access to the latest news and statements by Hamas. The application, which is available for both the iPhone and iPad, provides a selection of news from the PIC and also acts as a gateway to the

PIC website. However, it can only be accessed by searching iTunes for the term "Hamas" in Arabic.

The new branding of the Hamas website in various languages contributes to its "innocence." At a first glance, it would be very hard for a naïve user to notice the difference across the sites. Moreover, even those users who compare websites in different languages are very likely to find more uniformity than differences. Thus, Hamas uses different migration-of-brand tactics according to different target audiences.

Decision 5: Outsourcing and Strategy Partnerships

According to the e-marketing model, this decision is relevant to several dimensions of online presence: design or technology partnerships, promotion partnerships, reciprocal promotion, distribution partnerships, supplier partnerships, and legal advice. Hamas online has clearly used outsourcing, especially concerning servers that store its websites in different languages. Each site uses a different Internet protocol address, maintenance is done by a different company, and the site itself is located on a different server. Hamas goes further in safekeeping its Arabic Internet sites by posting them on three different servers.

Hamas' websites are hosted mostly in Western countries. The main provider is the United States, whose companies maintain 14 Hamas domains and provide network access to two domains, even though the US government has long designated Hamas as a terrorist organization. (Hamas also uses Canadian Internet service providers, even though the militant group is banned from operating in Canada.) One Hamas website, the online children's magazine *Al-Fateh* (see chapter 2), published at www.al-fateh.net, is hosted on US servers in Scottsdale, Arizona. The magazine has been published online since 2002, originally using the domain al-fateh.net, which is currently hosted in Moscow. Russia and Malaysia are the second-largest access providers to Hamas. Each country hosts five domains for Hamas. In addition, Hamas operates a network of online operatives: paid and volunteer members of the pro-Hamas Internet support group whose purpose is to infiltrate blogs, forums, and chatrooms with pro-Hamas propaganda.

A new and improved version of the website of Hamas' Izz al-Din al-Qassam Brigades was launched on June 8, 2008. The newly upgraded website was widely covered on many websites operated by Hamas and its supporters. According to an article published in the updated version of the website, on the day the new version went live, more than half a million users accessed the site. The upgraded website in both Arabic and English versions obtains its technical services from DataForce Contacts, a company based in Moscow.

In 2008, Hamas launched AqsaTube, its answer to the popular YouTube video streaming and sharing site. AqsaTube showed real-time videos about the Islamist movement and was, as the website announced in its introduction, "the first Palestinian website specializing in Islamic and jihad audiovisual productions." AqsaTube was taken offline but was replaced by a newer version, PalTube, at http://paltube.org/. It can be viewed in 20 different languages, provides access to thousands of Hamas videos, and offers options to upload and share videos.

Decision 6: Organizational Support

The e-marketing strategy emphasizes the organizational support that should accompany online marketing at all times. A special organizational system should provide steady support to the website and online marketing venues. The company can decide at any point to increase or decrease organizational support. The Hamas organization today gives the PIC its full support. The ideological and administrative regulation of the site is managed by the organization's offices in Damascus, where Hamas' political leadership resides. From there, Hamas controls the PIC websites, relaying orders through its Lebanese branch to the head Hamas activist Nizar Hasan Saliman Hassin. Hassin is listed as a contact on a number of the PIC websites, including the Arabic site. There is not sufficient information to make a clear picture of the organizational structure that Hamas built in order to support the PIC, but it is obvious that the organization has and does channel many resources to the operation of the PIC websites. This conclusion is drawn from the constant updating of the numerous websites; the uploading of visual, audio, and print material; the upgrading of their technology; and the application of up-to-date Internet platforms including Facebook, Twitter, YouTube, and Google Earth.

The sophistication of multiple websites in different languages, each catering to different target audiences with distinctive messages, reveals the existence and operation of production teams that speak different languages and translate and write articles. As described earlier, the importance of the differences between the sites and the lucidity of the message they convey may suggest a regulation function for each site and maybe even a central regulation system that continuously controls the ideas shown in each site, according to a clear, structured plan.

Decision 7: Building a Budget and Resource Allotment

The e-marketing model proposes two reference points for building a budget. The first reference point is the initial funding, taking into account expenditure factors and the cost of building and running the site. The second point considers different parameters, such as last year's Internet marketing budget, the percentage of company sales online, the percentage of total marketing budget, the reallocation of marketing expenses, the spending of other companies in the industry, the creation of an effective online presence, a graduated plan tied into measurable results, and the combination of these approaches. After examining the Hamas websites through this model, it is possible to understand how each of the expenditure components has an influence on the PIC websites:

- *Initial website investments:* The PIC website went live in Arabic on December 1997, followed in January 1998 by the English version. By 2002, the sites in other languages were up and running. It is safe to assume that establishing the website, supplying the necessary electronic tools, and putting it online were done with considerable financial investment approved by the Hamas leadership.
- *Continuous maintenance of the website:* The PIC websites operate like a news network. As pointed out in the previous decision point, the structure and role of each language website requires an investment in skilled manpower in different capacities, and providing such manpower requires budgets. Another section of maintenance of the websites is the protection of the sites against attacks from hostile groups and the financial cost if a website collapses.
- *Relaunching the website:* Every addition of a language to the site presents an additional cost. Furthermore, at the end of 2006 the PIC websites were enhanced; the entrance portal of all the sites was updated and all the sites became uniform in their appearance. It is evident that for the different marketing activities the organization initiated, including the renewing of the PIC websites, millions of dollars were raised in different Islamic countries.

The Council on Foreign Relations estimates Hamas's annual budget at $70 million (Laub 2014). The largest backer of Hamas is Saudi Arabia, with more than 50 percent of Hamas funds coming from that country, mainly through Islamic charity organizations. This funding continues despite Saudi pledges to stop funding groups such as Hamas that have used violence and its denouncements of Hamas' lack of unity with Fatah. According to the US State Department, Iran, another major funding source of Hamas, provided "weapons, training, and funding to Hamas and other Palestinian terrorist groups" (Department of State 2013, 196). Hamas maintains a strict separation of funds used for military operations and those used for political, social, or other activities. The majority of funds for military activities, around $3 million annually, comes from Iran, whereas funds from charity organizations or from Saudi Arabia and other Gulf states are used only for political and social activities.

It is almost impossible to estimate the benefits gained by Hamas through e-marketing. As a terror organization, the product being sold is an extreme ideology that is expressed through acts of terror that are almost completely planned, organized, and executed offline, in the real world. Furthermore, the PIC websites do not actively recruit activists, and registration is not required to read their contents. However, Hamas' continuous support of its online presence, its growing use of the Internet, and several statements made by Hamas' leaders on the importance of its websites are indicators of the organization's assessment of this platform's effectiveness.

Decision 8: Channel Structure Modifications

The model (Chaffey et al. 2000) presents various strategies designed to take advantage of changes in marketplace structure. These options are (1) disintermediation (sell direct), (2) creating a new online intermediary (countermediation), and (3) partnering with new online or existing intermediaries. To achieve strategic Internet marketing goals, organizations also have to plan for technology integration with customers' and suppliers' systems. However, an even more vexing question is how to manage the channel conflicts involved with new developments and new environment.

The PIC websites certainly demonstrate changes and modifications based on flexibility of online messaging. Hamas is reacting rapidly to events, posting online announcements and references to occurring developments and processes. The websites change contents mostly on a daily basis, and refer not only to local events but also to global developments (e.g., the election of Barack Obama). Moreover, as described above, Hamas websites change their formats and design: adding new features, posting new links, applying new online technologies, and adopting recent innovations in cyberspace—from

Facebook to YouTube. However, it should be noted that all these changes are not original innovations but are based on a "copycat" approach.

The application of the e-marketing model developed by Chaffey and his colleagues (2000) to the case of Hamas websites is an illustrative attempt to test the usefulness of commercial, business-world–based models to the case of online terrorism. The model, based on eight decision points, was designed to outline a strategy for businesses that want to use the Internet as a marketing platform. They claim that e-marketing is becoming a major apparatus for modern promotion of goods and services. The main function of the proposed model is its systematic outline of key stages in building, maintaining, and changing marketing-oriented websites. Although the model was developed for commercial purposes, our study demonstrates how it could be used as an analytic framework to study terrorist websites. As shown, most of the decision points in the model were relevant to the PIC websites. Obviously, these findings do not in any way suggest that the model was the guiding manual for the decision makers in Hamas, but using the model to understand Hamas' presence on the Internet may tell us how the group's decisionmakers perceive their functions in the virtual world (Mozes and Weimann 2010).

Chapter 5

Debates Online

Terrorists are using the Internet for various purposes. Most of the attempts to monitor and study terrorist presence online have focused on the practical and communicative uses of this channel by modern terrorists. Yet not much attention has been paid to the use of the Internet as a medium for terrorist debates and disputes. This chapter presents this less-noticed facet of Internet terrorism by providing examples of virtual debates among and within terrorist groups. The analysis of the online controversies, disputes, and debates may tell us a lot about the mindsets of terrorists and their motivations, doubts, and fears. In many ways, it opens a window to a world we know little about. It may also serve counterterrorism efforts: by learning about the inner cleavages and debates within terrorist groups, one can find practical ways to support the voices against terror, broaden gaps within these dangerous communities, and channel online discourse to nonviolent forms of action.

This chapter updates and expands on material originally published in "Virtual Disputes: The Use of the Internet for Terrorist Debates" (Weimann 2006c).

Debates within Terrorist Groups

Al-Qaeda's Inner Debates

There are numerous cleavages and rifts within terrorist groups. Some of these are revealed when studying online communications. One example is the case of a rivalry within al-Qaeda. It appears that from within this loosely knit organization emerged a group of young Saudi Islamists who wished to play a more important role. As Reuven Paz (2003, 2) noted, "Many of them were students and disciples of the older groups of Wahhabi clerics and scholars who could not come to terms with the American presence on Saudi soil. In recent years they radicalized their positions and began backing up the positions of Qa'idat al-Jihad, including political violence against the United States, Western culture and, in recent years, the Saudi royal regime, while providing Islamic legitimacy for these actions." In 2002, these groups began to issue new electronic pamphlets through the websites of their supporters, using a new name: Qa'idat al-Jihad (the Jihad Base). The severe conflict within the Saudi jihadist groups led to the online publication of Saudi scholar Abu Jandal al-Azdi's 460-page book, *Osama bin Laden: Mujaddid al-zaman wa-qahir al-Amrikan* (Osama bin Laden: The reformer of our times and defeater of the Americans). The book attempted to raise bin Laden to a new level of a major reformer or reviver, a status bestowed on only a select few scholars.

On December 7, 2006, a jihadist website posted a letter addressed to the "Al Qaeda Organization in the Land of the Two Rivers" (i.e., Iraq) titled "The Solid Structure." The posting criticized al-Qaeda's strategy in Iraq and proposed an alternative approach to the unity of jihadist groups in Iraq. The author, Abu-Hamzah al-Ansari, introduced his letter as a response to the call of Abu-Hamzah al-Muhajir, a former active combatant of al-Qaeda and a senior aide to the late Abu Musab al-Zarqawi, former leader of Al-Qaeda in Iraq. This call instructed jihadist groups in Iraq to pledge allegiance to the Islamic state in Iraq, and it urged al-Qaeda members in Iraq to "show courage by listening to the other views." Al-Ansari urged all mujahideen "to be brave by discussing their vital issues openly and frankly." He proposed that mujahideen emulate "the courage of the West led by America in its practice of self-criticism." He said that while it is true that the Americans practice "hypocrisy and denial of justice in their dealings with the rest of the world," they are "quick to point out problems, criticize, and acknowledge mistakes among themselves." Al-Ansari questioned al-Qaeda's call for allegiance and what he insinuated was a premature decision to announce an Islamic state in Iraq. He identified himself as a member of a small jihadist group, and added that his letter "should be understood as a letter from a Muslim on behalf of a small group of Muslims exercising their religious right to express their views and disagreement, rather than a letter from a small group to a larger group." Al-Ansari criticized what he saw as al-Qaeda's efforts to gain legitimacy at the expense of the Iraqi people's respect for their history. He rejected what he called al-Qaeda's incessant attacks on the Ba'thist regime and sought to point out "honorable achievements" in the history of Iraq. On December 8, the following day, jihadist websites posted reactions from the Al-Buraq website; forum participants criticized al-Ansari's letter.

Al-Qaeda's presence in Iraq involves many inner debates within jihadi groups. In the absence of dominant personalities such as Abu Musab al-Zarqawi (who was killed by US forces in 2006) and the anonymity of his successors, some of the other jihadi or Islamic groups of the Sunni insurgency allowed themselves to either criticize al-Qaeda or enter into clashes with its members, which were followed by some violent attacks by al-Qaeda. As Paz (2003, 8) revealed, the absence of dominant scholars created a big vacuum: "This vacuum left the door open to more public criticism and debates within the Jihadi-Salafists on one hand, and makes them more vulnerable to outside criticism and attacks by Saudi anti-Takfiri Salafists and affiliated scholars, on the other. The whole process of criticism and debates is done in public, on the Internet."

The virtual disputes within al-Qaeda also relate to disagreements regarding operations. For example, in 2012, Abu Zubeir Adil al-Abab, a sharia (Islamic law) official in Al-Qaeda in the Arabian Peninsula (AQAP), addressed a message to fellow scholar Abu Baseer al-Tartusi concerning his remark that Ansar al-Sharia's suicide bombings against the Yemeni army have no legal basis in Islamic law. The message, "Would You Leave Me to Our Sheikh Abu Baseer al-Tartusi?!," was published by al-Tawhid Media, a group affiliated with the Pulpit of Tawhid and Jihad website, and posted on their site and other jihadist forums on April 13, 2012. In his message, al-Abab defended the group's use of suicide bombings, explaining that they are primarily used as a last resort to hit an enemy target or to "achieve some greater good," and gave several examples in the past year of how the tactic proved beneficial to the fighters. Regarding al-Tartusi's opposition to strikes against soldiers who stood with the Yemeni revolution, "Abab told him that the Yemeni army is no longer as he remembers it." Al-Abab claimed that the current incarnation of the army does not support the revolution, and in instances where fighters have captured soldiers, they have released those who repented, even though those soldiers had killed their colleagues. He also remarked that the soldiers should realize the nature of the war and leave their service, further stating: "The issue of the soldier has become larger than before, and as I said, the soldier has become a victim of the leanings of the parties and groups that are ruled by the West and Iran. The battle has become clearer in the eyes of those soldiers, especially in the last year."

A more recent dispute is the controversy between al-Qaeda's branches in Iraq and Syria. In June 2013, an online audio message surfaced from Abu Bakr al-Baghdadi, the head of al-Qaeda's branch in Iraq—the Islamic State of Iraq (ISI)—defying Ayman al-Zawahiri's command that the group that al-Baghdadi had created—the Islamic State of Iraq and the Levant (ISIL)—be disbanded. Jihadists' hopes for the settling of the conflict vanished and they became confused, with many doubting the authenticity of al-Baghdadi's message. However, in the days after the message's release (as the administrators were unable to suppress the high volume of posts about the subject), the consensus was that the audio was indeed real. The authenticity of the audio files was ultimately confirmed by administrators of the Shumukh al-Islam forum in a statement posted five days later, on June 19, thus bringing to the forefront issues about al-Qaeda and leadership, jihad in Iraq versus Syria, and the current track of the global jihadi movement. This dispute is in fact a power struggle, with one side demanding control over al-Qaeda's affairs in Syria and Iraq, and the other desiring independence from the Iraqi side—as well as reporting directly to al-Qaeda and al-Zawahiri (Katz and Raisman 2013).

In January 2014, al-Qaeda's central leadership declared the split with the faction known as the ISIL. The announcement issued on Islamic militant websites appeared to be a move by al-Zawahiri to reassert the terror network's prominence in the jihad movement across the Middle East amid the mushrooming of multiple extremist groups over the past three years. The declaration prompted harsh reactions: On militant websites, ISIL supporters lashed out at al-Qaeda's leadership. "God as my witness, al-Qaeda did not do right by this mujaheed group. Instead, it stood with its enemies," one supporter with the username Muslim2000 wrote. A spokesman for the Islamic Front vowed that it will continue battling the Islamic State: "The Islamic State is now without cover or co-sponsor. It has been totally stripped after al-Qaida and the people abandoned it" (Youssef and Keath 2014).

The complex situation in Syria has caused an inner split within al-Qaeda–related groups. In an unexpected and unprecedented turn of events, al-Qaeda members and jihadists from all over the world who had embraced the ideology of global jihad are now doubting the group's leader, Ayman al-Zawahiri, and calling for his removal. As Katz and Raisman (2014, 1), reported, "While the Syrian Jihad has been of paramount concern to world governments, with hundreds of foreign fighters pouring in to participate in the fighting, the country has also been an arena for internal strife and bloody battles between the al-Qaeda-affiliated al-Nusra Front and al-Qaeda's former branch, the Islamic State in Iraq and the Levant (ISIL). Members of al-Qaeda and jihadists together blame Zawahiri for mismanaging the conflict in Syria and enflaming sedition, and advocate new leadership." Under al-Zawahiri's command, al-Qaeda assigned the al-Nusra Front to be its arm in Syria and disavowed the ISIL, telling jihadists that al-Qaeda does not acknowledge its founding and rejects its activities. Representing al-Qaeda's core leadership, Adam Gadahn (also known as Azzam the American) posted online messages arguing that jihadi factions in Syria blamed a "group that is known for its extreme nature and radical behavior"—statements that infuriated pro-ISIL supporters, who believed that Gadahn was referring to the ISIL. When al-Zawahiri himself made similar accusations in a speech released online in April 2014, some jihadists quickly attacked him, questioning his wisdom and demanding his removal, and did so on al-Qaeda–affiliated password-protected forums. According to Katz and Raisman (2014), the assault on al-Zawahiri became so fierce that administrators of the top-tier jihadi forum Shumukh al-Islam deleted all the posts in the discussion thread for the speech ("Eulogy for the Martyr of Sedition Abu Khalid al-Suri") and locked it. A few hours later, the forum went down for "maintenance."

Hezbollah's Inner Debates

Hezbollah's military involvement in the fighting in Syria alongside the Syrian regime has caused a rift within the organization and among its supporters in the Lebanese Shiite population, who form the main base of its popular, political, and military might in the country (Picali 2013a). According to E. B. Picali (2013b), the criticism and reservations expressed by the pro-Hezbollah Shiites regarding the organization's fighting in Syria are motivated by two main reasons. First, the fighting in Syria is seen as an act that contradicts the organization's claim that its weapons are only meant to fight Israel. Second, there are serious concerns that involvement in Syria could drag Lebanon into a Shiite-Sunni sectarian war, at a high cost to the

Shiite population. It should be mentioned that, back in October 2012, the British *Daily Telegraph* reported on disagreements within the organization between Hezbollah's military wing, which claimed that it supported the involvement in Syria, and its political wing, which rejected this involvement (Meo, Sherlock, and Malouf 2012). Initially, reports on criticism among the Shiite population were anonymous, but later on, and as Hezbollah casualties in Syria increased, its Shiite supporters increasingly began to make their opinions known openly in the Lebanese press. The reports on bitter families of fighters appeared mainly in anti-Hezbollah media, but occasionally also appeared in media known for supporting Syria and Hezbollah. Picali (2013a) argues that the Hezbollah is worried about losing popular support from the Shiite public—the base of its political and military power—and is therefore making efforts in two areas: convincing the Shiite public that its fighting in Syria is justified, and silencing—or at least minimizing—the criticism.

Debates among Terrorist Groups

Hamas versus al-Qaeda

A lingering open dispute between Hamas and al-Qaeda started on the Hamas website (Paz 2004). The Palestinian Hamas movement and al-Qaeda have a complex relationship. Al-Qaeda and the affiliated Salafi-Jihadi or Tawhidi-Jihadi groups are trying to manipulate the Palestinian problem and the Israeli-Palestinian conflict, while Hamas uses the same issues for its own propaganda, recruitment, and fund-raising. Al-Qaeda and its affiliates "also adopted from Hamas the *modus operandi* of suicide attacks and much of the Islamic justification for them. On the other hand, Hamas has so far tended to reject attempts to create organizational links with the global jihad, or to expand its terrorist activities beyond Israel and the Palestinian territories" (Paz 2004, 1). On May 30, 2004, Hamas published on its official website the following press release related to the May 29 Khobar attack in Saudi Arabia, where 22 people, most whom were foreigners, were killed in an attack on Saudi oil industry installations by an al-Qaeda–related group:

> The Islamic Resistance Movement Hamas declares its severe condem-
> nation and sorrow for the criminal attack that occurred yesterday night
> in one of the complex of buildings in the town of Khobar in the brother
> kingdom of Saudi Arabia, which caused the death of dozens of civilians
> and innocent people. While we reject this kind of attacks [*sic*], we wish

to emphasize that they harm the security and peace of our countries, and the national and Islamic interests. Therefore, we call those responsible for these attacks to stop them, and preserve the interests and security of their country and nation, especially while our nation is facing external threats and challenges. (Quoted in Paz 2004, 3)

This Hamas online statement openly criticizes the al-Qaeda attacks for their harmful implications.

Hamas has also used the Internet to respond to its critics. In January 2004, Hamas posted photos on its Internet site of Reem Raiyshi, its first female suicide bomber, posing with her two young children. Raiyshi blew herself up at a border crossing between the Gaza Strip and Israel, killing four Israelis. The terror group's posting was an attempt to answer Palestinian critics who condemned sending a mother on a suicide mission. In the pictures, Raiyshi is posing in camouflage dress and the Hamas headband, holding an assault rifle, and her three-year-old son holds a mortar shell. Sheik Saed Seyam, a Hamas leader, claimed, "This picture shows the outrage the Palestinians have reached. This scene should urge people to ask themselves what motivates women, who are known for their attachment to their children and families, to leave them forever." When the sheikh was asked why Hamas had sent a mother to her death, he answered that there are many volunteers who want to carry out attacks: "Some of them cry [to be chosen], which makes the military leadership submissive to their pressure for this honor" (quoted in "Happy Snaps of a Suicide Bomber" 2004).

In an online audiotape released in 2007, Ayman al-Zawahiri, then the deputy of Osama bin Laden, attacked the Hamas movement, blaming its leadership for accepting the establishment of a Palestinian unity government and thus "selling Palestine to the Jews" in exchange for government seats. Accusing the Hamas government of being helpless and dependent on outside elements, such as Israel and Egypt, al-Zawahiri said that Ismail Haniya, the prime minister of the Hamas government, could not even enter his own house without Egyptian mediation with the Israeli defense minister. This attack caused anger among Hamas leaders and operatives. Hamas leaders were quick to respond to what they referred to as "baseless claims," and stressed that despite the signing of the Mecca Agreement[1] the movement had

---

[1] The Mecca Agreement, signed on February 8, 2007, ended the military conflict between Fatah and Hamas in the Gaza Strip that had followed Hamas's electoral victory in the Palestinian legislative elections in January 2006.

not abandoned its principles. On the official Hamas website, they published an official statement titled, "You have misunderstood, Dr. Al-Zawahiri, and you failed in your statement." According to the announcement, Hamas still adheres to the path of jihad and resistance (i.e., violence and terrorism) and will continue to do so, sacrificing martyrs "until not a single trace of occupation remains in Palestine." Nonetheless, in a 21-minute speech titled "Palestine Is Our Concern and the Concern of Every Muslim," issued by the al-Qaeda multimedia production arm as-Sahab on March 11, 2007, al-Zawahiri argued that "the leadership of Hamas surrenders to the Jews most of Palestine." Al-Zawahiri believed that Hamas has "sunk in the swamp of surrender," and called upon Muslims to reject politics and engage in jihad against the enemy (Intelligence and Terrorism Information Center 2007).

The Global Islamic Media Front (GIMF) and the al-Fallujah jihadist forum, both al-Qaeda outlets, launched a media campaign on October 7, 2009, to expose Hamas's involvement in the death of radical sheikh Abu al-Noor al-Maqdisi. Al-Maqdisi, the spiritual leader of the al-Qaeda–inspired Islamist organization Jund Ansar Allah, was killed on August 15, 2009, during a Hamas raid on Ibn Taymiyya Mosque in Rafah, Gaza. The campaign, titled "Media Invasion of the Martyr Imam Abu al-Noor al-Maqdisi," informed Muslims that the money they donate to Hamas is "bullets in the chests of Muslims," and that the Hamas of today is not the same as that of yesterday. In addition, the campaign published al-Maqdisi's speeches. GIMF stated: "Let them know we have not forgotten the spilling of the blood of the believers. Tell them we are proud people who do not get weak or soft. We will march on to reveal the disgraceful falsehood." Campaigns such as this one highlight the political and social rifts between Hamas and al-Qaeda, and demonstrate the extent to which both groups will turn to online platforms to promote their own views and denigrate those of their opponents.

Hezbollah versus Other Islamist Groups

When Israel and the Lebanese Shiite group Hezbollah exchanged prisoners on January 29, 2004, Hezbollah and its leader, Hasan Nasrallah, enjoyed a glorious boost to their image, since the Israelis had had to release 430 Arab prisoners in exchange for the bodies of three Israeli soldiers and one kidnapped Israeli businessman. This created criticism and resentment among several Arab groups and among groups opposed to the Hezbollah. The most serious criticism of Hezbollah came from the Saudi Jihadi-Salafi elements that support Qa'idat al-Jihad (also known as the Al-Qaeda Group of Jihad in

Iraq). Hezbollah has never been popular among the Salafi adherents of global jihad, given the latter's fundamental hatred of the Shia. Since the beginning of the attempts to establish a new government in Iraq in 2003, Salafi websites and Internet forums have stepped up their attacks against the Lebanese Hezbollah. The most significant verbal attack was the labeling of Hezbollah with the Salafi term *Hizb al-Shaytan* (the Party of the Devil). One of the primary Salafi online forums against non-Sunnis (mainly against Shiites), al-Difa an al-Sunnah (Defense of the Sunna), led the attacks with numerous writings, several of which were by important Islamic writers. Their website presents articles such as, "The actions of the murderer so-called Shiite Mahdi," "The crimes and betrayals of the Shi'is throughout history," "Meetings between Shiite clerics and Jews and Christians," and "The scandals of Shiite clerics and religious authorities." In this way, younger Islamists use jihadi forums to launch personal attacks on Nasrallah in person; one poster in the al-Erhap jihadi forum titled his attack, "Hasan Nasrallah, leader of Hezbollah—the most famous and corrupt traitor in the history of the nation" (Weimann 2006b, 142–43).

The al-Qaeda affiliate Al-Qaeda in the Islamic Maghreb (AQIM) launched its online anti-Hezbollah campaign in 2013. In a series of messages issued to the Twitter account of their al-Andalus Media Foundation, AQIM blamed Hezbollah for the August 23, 2013, bombings in Tripoli, Lebanon. The tweets, posted the same day as the bombings, offered condolences to the Sunni Muslims of Tripoli and reported that the group knew "with certainty" that Hezbollah was responsible for the attack, which left more than 50 people dead and more than 500 wounded. The posts stated that "heroes" in Syria would not let the event pass without taking revenge against Hezbollah at the appropriate time (SITE Monitoring Service 2013b).

The Shiite-Sunni Divide

One of the most vocal and even aggressive disputes within the Muslim world is the Shiite-Sunni conflict. This rift, which also divides radical Islamists and terrorist groups, is further revealed in their online animosity. On January 18, 2007, members of the password-protected jihadi forum Mohajroon called upon the Sunni people in Iraq to confront the alleged Shiite aggression against their people and mosques. Included in the Mohajroon forum postings were manuals for making explosives, and suggestions for placing bombs and positioning soldiers. The posts suggested that every Sunni, regardless of sex or age, should take up weapons and organize into groups for dispersal in strategic locations to fight the Shiite militias. Others, however, believed that the women and children should be protected as the men contribute to the fighting. Snipers were stressed as mandatory for the confrontations. Another member provided instructions for engaging in hand-to-hand combat with the enemy—even using a newspaper as a weapon—and argued that the Shiites are weak in this realm: "Our enemies neglected the physical training and depended on the power of the modern weapons." In another online posting, a group called "Abu-Mus'ab Brigade" posted a message to several jihadist forums in which the group's "supreme command" announced that they will be joining in the "great battle of Baghdad" and would target the "vital Shiite news agencies."

The Sunni-Shiite divide also manifested itself in the problematic relationship between the Shiites of Hezbollah and the Sunni of al-Qaeda. In January 2007, the message "Nasrallah Is Setting Lebanon on Fire," a message attributed to al-Qaeda strategist Abu Musab al-Suri, surfaced online. It was circulated among jihadist forums, including the password-protected al-Qaeda–affiliated al-Hesbah Network. Ridiculing Hezbollah leader Hassan Nasrallah as "Nasr-Ellat"—a reference to a pre-Islamic pagan goddess—the text explained that jihad is an incumbent duty upon Sunni Muslims, exacerbated by the Shiites and Hezbollah's attempts to assert influence in the region over the people. Al-Suri urged, "Here he is, the fake Nasr-Ellat declaring his failure, and the failure of his satanic party. And here they are, his followers, the Cross-worshipers who have a history of slaughtering the Lebanese people, went out to the streets, set the green trees and the cars on fire, destroyed the storefronts, and threatened to kill the people. . . . Youth, arm yourselves and get ready to confront the Shiite death squads. Support the Sunni Mujahideen. Do not be afraid of the Shiites, because they are cowards."

The Sunni-Shiite conflict has had several manifestations, one of which is cyberwar: Hundreds of Sunni and Shiite websites, including sites of clerics, newspapers, and government ministries, have been hacked and defaced with offensive messages and images (Azuri 2008). The Sunni-Shiite cyberwar started in 2007 when a group of Sunni hackers, calling itself "XP Group," threatened to attack all Shiite websites, and proceeded to hack some 120 Shiite sites. (Most of the hacked sites reemerged within weeks.) At this point, representatives of the targeted sites, headed by Shiite sheikh Hassan al-Saffar, filed a lawsuit in Saudi Arabia against a member of the XP Group named Na'if al-Ghamdi. According to recent reports on Sunni forums, al-Ghamdi has been arrested and has disclosed the names of 17 other hackers operating in Arab countries. In response to these incidents, a number of Sunni hacker groups pledged to retaliate against Shiite sites. Among the hacker groups vowing vengeance were two groups called Shabab al-Salafiyin and al-Ayyoubiyoun. The latter declared on various forums that the war against Shiite sites was a form of jihad that brought one closer to Allah. The threats were realized in August 2008, when a group of Egyptian and Saudi hackers attacked the Shiite sites Fatimid Egypt, Egyptian Shia, and others. This prompted a Shiite group, Shabab al-Shia, to threaten further attacks against Sunni sites. This cyberwar continues to the time of this writing, in tandem with the Sunni-backed ISIS's physical attacks on Shiites in Syria and Iraq.

What Is Not Debated?

In light of the varied debates taking place in the abovementioned corners of the Internet, it is important also to learn about the issues that are not debated by terrorists. For example, the Arab media has extensively debated the use of suicide action, but this debate has not occurred among the organizations applying this method. Following the wave of suicide bombings in Israel, Iraq, Bali, and other arenas in the late 1990s and early 2000s, many plunged into a hot debate concerning the religious, political, and moral legitimacy of this form of operation (MEMRI Jihad & Terrorism Studies Project 2001).[2] The mufti of Saudi Arabia, Sheik Abd al-Aziz bin Abdallah al-Sheik, argued, "I am not aware of anything in the religious law regarding killing oneself in the heart of the enemy['s ranks], or what is called 'suicide.' This is not a part of Jihad, and I fear that it is merely killing oneself. Although the Koran permits and even demands the killing of the enemy, this must be done in ways that do not contradict the *Shari'a* [Islamic law]."[3] However, among terrorist organizations and their leaders, the legitimacy of suicide actions is not debated or criticized. Sheik Hamed al-Bitawi, head of the Hamas-affiliated Palestinian Islamic Scholars Association, stated that, according to Islamic law, "Jihad is a collective duty [*fardh jifaya*]. . . . However, if infidels conquer even an inch of the Muslims' land, as happened with the occupation of

---

[2] This issue is important, since the Quran forbids suicide, and the fatwas (Islamic legal rulings) are the only way to legitimize suicide attacks by turning them into forms of self-sacrifice or death in combat.

[3] Published in *Al-Sharq Al-Awsat* (London), April 21, 2001.

Palestine by the Jews, then Jihad becomes an individual duty [*fardh 'ayn*]," and therefore, suicide attacks are permissible.[4] Abd al-Aziz al-Rantisi, one of the leaders of Hamas later killed by Israeli forces in 2004, joined Sheik al-Bitawi and explained that "suicide depends on volition. If the martyr intends to kill himself, because he is tired of life—it is suicide. However, if he wants to sacrifice his soul in order to strike the enemy and to be rewarded by Allah—he is considered a martyr [rather than someone who committed suicide]. We have no doubt that those carrying out these operations are martyrs."[5] Al-Bitawi and al-Rantisi based their statements on a fatwa by Sheik Yusuf al-Qaradawi, one of the leaders of Egypt's Muslim Brotherhood and a central religious authority in Sunni Islam. Sheik al-Qaradawi published a fatwa stating that suicide attacks are allowed according to Islamic law. In a February 2001 interview in the Egyptian newspaper *Al-Ahram Al-Arabi*, al-Qaradawi explained his ruling:

> He who commits suicide kills himself for his own benefit, while he who commits martyrdom sacrifices himself for the sake of his religion and his nation. While someone who commits suicide has lost hope with himself and with the spirit of Allah, the *Mujahid* is full of hope with regard to Allah's spirit and mercy. He fights his enemy and the enemy of Allah with this new weapon, which destiny has put in the hands of the weak, so that they would fight against the evil of the strong and arrogant.[6]

Thus far, argues Paz (2005, 78), the main *modus operandi* of the global jihad (and mainly Qa'idat al-Jihad) has been suicide or martyrdom operations: "Martyrdom attacks are not only a tactical tool of terrorism; they have also played a central role in the indoctrination of al-Qaeda recruits. . . . This idea of self-sacrifice has since been reinforced as the phenomenon of operations has spread across many parts of the world, not to mention by the worldwide increase of support of Muslim publics for the suicide attacks against civilians in Israel. It is significant to note that this method, which was once controversial among Islamic clerics and scholars, enjoys growing support within religious and political communities alike. Thus far, in fact, it seems that radical Islam's focus has been not on mass-killings, but primarily on

---

[4] Published in *Al-Hayat* (London-Beirut), April 25, 2001.
[5] Ibid.
[6] Published in *Al-Ahram Al-Arabi* (Egypt), February 3, 2001.

self-sacrifice and on the proliferation of its attacks to different regions and places across the globe."

The focus on personal martyrdom and suicide attacks among the groups that adhere to the culture of global jihad—as well as groups with more local and national aspirations, such as the Chechen Islamists and the Arab volunteers there, the Kashmiri groups, the Kurdish Ansar al-Islam, or the Palestinian Hamas and the Palestinian Islamic Jihad—might explain the absence of any debate or controversy over this tactic among the members and supporters of these groups. Similarly, Paz (2005) notes the lack of dispute over the use of weapons of mass destruction (WMD) by terrorists. The absence of a discussion over WMD in terrorist websites and forums, in online discussions or postings, is significant. In recent years, Qa'idat al-Jihad and affiliated groups have issued only a few pronouncements in which they threatened to use WMD. The first direct reference appeared on December 26, 2002, when Abu Shihab al-Kandahari, the then moderator of the Islamist Internet forum al-mojahedoon.net, published a short article titled "Nuclear War Is the Solution for the Destruction of the United States."

On May 21, 2003, the Saudi Shaykh Naser bin Hamad al-Fahd published the first fatwa on the use of WMD. The ruling was posted online by the Global Islamic Media Center, al-Qaeda's formal mouthpiece. The author, who supports the culture of global jihad and the militant struggle against the West, has published numerous militant books and articles that the followers of global jihad regard as religious rulings that legitimize the war on the West and the forms of operation used by the fighters. Shaykh al-Fahd has been at the forefront of a new effort to rethink the strategy of asymmetric warfare shared by many Islamists. After the 9/11 attacks, he argued: "If the Americans are using F-15 or Tornados [and they are allowed to do so], then if the Mujahideen used Boeing or Air Bus are they not allowed?" Al-Fahd has repeatedly used such analogy with the West to provide legitimacy for jihadists' use of WMD. He even cited the Prophet Muhammad in the Hadith: "Allah has ordered you to do everything perfectly. Hence, if you kill, do it perfectly, and if you slaughter, do it perfectly. Everyone should sharpen his blade and ease his slaughter." In al-Fahd's view, this means that jihadists should maximize their abilities by every means possible, including WMD. It is interesting to note that this fatwa was not intended to trigger any debates or lead to any questioning. Moreover, combined with similar declarations issued by Ayman al-Zawahiri, jihadists seem to accept this ruling. As Paz (2005, 78) noted, "Shaykh al-Fahd's ruling was not accompanied by any dispute or discussion. In fact, those who follow the many radical Jihadi websites, forums, and chatrooms—the main arena of the discourse for radical Islamists—may well have been surprised by the absence of any coherent debate on WMD of any kind among Islamists. In some cases, Islamists expressed their hopes and desires that al-Qaeda use chemical, biological, radiological or nuclear weapons (CBRN) against the West."

Numerous online terrorist manuals contain plans and instructions for the use of various weapons, but only a handful refer to the use of WMD. In March 2005, a terrorist group published a do-it-yourself plan to make a "dirty bomb" (a bomb that combines conventional explosives and a radiological weapon) on its Internet site. Alma'sadah al-Jihadiah, the site that published the plan, is run by a group whose aim is to promote and propagate terror activities. In October 2005, a "New Jihad Encyclopedia" was posted online, after its publication was first announced on the forum at alfirdaws. org and copied to the al-farouq.com site, both leading jihadi online forums. The "Encyclopedia," which contains nine lessons in approximately 80 pages in Arabic, was published under the title "The Nuclear Bomb of Jihad and the Way to Enrich Uranium" and was presented as "a gift to the commander of the jihad fighters, Sheikh Osama bin Laden, for the purpose of jihad for the sake of Allah."

Mustafa Sit-Maryam, better known as the former leading al-Qaeda trainer and scholar Abu Mus'ab al-Suri, published in December 2004 two significant documents calling for a new organization of global jihad: "The Islamist Global Resistance." The documents, published on al-Suri's website, discuss the importance of using WMD against the United States (Paz 2010). Al-Suri uses the American nuclear bombing of Japan in World War II to justify al-Qaeda's potential use of WMD, and even cites the example of President Harry S. Truman, who said that America's use of such bombs against Japan both shortened the world war and was fitting retaliation for the barbaric behavior of the Japanese. According to al-Suri, the United States today is no different from Japan in World War II, and therefore deserves to be targeted with WMD.

A Prospect for Counterterrorism

Terrorists use the Internet as a channel of communication among groups, within groups, from leaders to operatives, among followers and sympathizers, and among different currents within the same movement. In the war on terrorism, driving wedges between hardcore terrorists and their circle of sympathizers and fronts is essential to success. The 9/11 Commission Report (see chapter 1) noted the potential for this form of terrorist disruption by stating that "the commission claims that it is possible, through the use of public diplomacy, to drive a wedge between those moderate Muslims and the violent terrorists or insurgents" (National Commission on Terrorist Attacks upon the United States 2004, 376).

Samuel Helfont's book *The Sunni Divide: Understanding Politics and Terrorism in the Arab Middle East*, describes the inner split within the Sunni Muslims (Helfont 2009). In his conclusion, Helfont argues that the great divide in Islam is not the Shia-Sunni conflict, as numbers of Arab leaders have voiced, but rather the division within the Sunni Islamist community. To make his point, he narrows the divisions to the philosophical and political differences between two radical Islamist movements, the Muslim Brotherhood and the Wahhabists (or Salafists). This is an important issue because, as Helfont recommends, there are ways in which Western antiterrorism strategists can drive a wedge between these groups and pit one against the other.

The potential for using the revealed inner splits is crucial in the case of unstructured, decentralized al-Qaeda, which is struggling to project an image of robust operational capabilities. As suggested by James Forest (2012, 5),

In order for Al-Qaeda to convince its intended audiences of its status as a vanguard of jihadists defending the global Muslim community, it must establish and sustain a perception of integrity, worthy of trust and respect. The words and actions of those who have answered the call to jihad have, at times, created difficulties in shaping these kinds of perceptions. As many of us will recognize in our personal experiences, trust is much easier to break than to build. On an organizational level, Al-Qaeda has a significant challenge with regard to building and maintaining trust within the Muslim community. . . . The case of Al-Qaeda represents an example of influence warfare. Counterterrorism efforts should seek to . . . diminish the group's influence capabilities, and drive wedges in the solidarity of the movement that can help undermine and discredit its mobilizing ideology.

Finally, in his statement before the United Nations Security Council Counter-Terrorism Committee on May 24, 2013, Frank J. Cilluffo presented his paper, "Countering Use of the Internet for Terrorist Purposes." He also highlighted the notion of driving wedges between groups and factions, using inner debates and splits to weaken terrorist efforts:

> Brokering infighting between al Qaeda, its affiliates and the broader jihadi orbit in which they reside will damage violent extremists' capability to propagate their message and organize operations both at home and abroad. . . . We all could and should do more to drive wedges and foment distrust, including by exploiting points of conflict between local interests and the larger global aims of al Qaeda (and its ilk), and encouraging even more defectors (Cilluffo 2013, 6).

In many ways, studying terrorist debates online opens window to this little-known world. By learning its inner cleavages and debates, one can find practical ways to support the voices against terror, broaden gaps within terrorist communities, and channel their general discourse toward more non-violent forms of action.

Chapter 6

Online Fatwas

In 1989, the term "fatwa" became globally known when Iranian ayatollah Ruhollah Khomeini issued a fatwa that ordered Muslims to kill author Salman Rushdie, whose novel *The Satanic Verses* was deemed to be blasphemous to Islam. Few fatwas are as specific and extreme as the one pronounced on Rushdie, but jihadists continue to use them to promote similar forms of terrorist activity. Today, the Internet has become a useful platform for posting fatwas and interpretations of fatwas. This chapter describes the use of jihadist fatwas, especially online fatwas, as a major instrument in bridging the current wave of terrorism and religion. The analysis illustrates how cyber-fatwas are related to key issues in promoting terrorism, used to justify such actions as the use of suicide terrorism; the killing of innocents, women and children, and other Muslims; or the use of various weapons (including weapons of mass destruction and cyberterrorism). There are two

This chapter expands on material originally published in "Cyber-*Fatwas* and Terrorism" (Weimann 2011b).

implications of the trends documented here. First, the analysis of the online fatwas and the fatwa wars may tell us about terrorist motivations, doubts, and fears; second, it may guide counterterrorism campaigns.

The Emergence of Cyber-Fatwas

A fatwa is an Islamic religious ruling, a scholarly opinion on a matter of Islamic law. Because there is no central Islamic priesthood, there is also no unanimously accepted method to determine who can issue a fatwa and who cannot, leading some Islamic scholars to complain that too many people feel qualified to issue fatwas (Bar 2006a). In Sunni Islam, fatwas are non-binding and therefore not obligatory, whereas in Shia Islam an individual could consider the fatwa to be binding, depending on his or her relation to the scholar. The person who issues a fatwa is called, in that respect, a mufti (i.e., one who issues a fatwa). As described by Shmuel Bar (2006b, 1),

> The mechanism by which the scholar brings the principles of *shari'ah* to bear in the practical world is *fiqh*—Islamic jurisprudence, and its product is the *fatwa*—a written legal opinion or ruling on a specific subject, which dispels uncertainty and shows the clear path for behavior on the chosen subject. A *fatwa* can only be given by a scholar with wide enough knowledge of *shari'ia* to be considered a mufti. The classic fatwa consists of a question (*istifta'*), posed by a petitioner (*mustafti* pl. *mustatifun*), and a response (*jawab*). A fatwa must be based on the sources (*usul*) of *fiqh*: these include the Qur'an, the Sunna, logical analogy (*qiyas*) and consensus of the "*ulama*." However, most *fatwas* make little use of these tools and instead very often cite precedents from decisions by the *mujtahidun* of early Islam and the codex of existing fatwas.

A fatwa may concern many aspects of individual life, such as social norms, religion, war, peace, jihad, and politics. Millions of fatwas have been issued over the 1,400-year history of Islam; most deal with issues faced by Muslims in their daily lives, such as the customs of marriage, financial affairs, female circumcision, or moral questions. They are usually issued in response to questions by ordinary Muslims. The assassination of Egyptian president Anwar Sadat in 1981 was attributed to a fatwa issued against him by Egyptian Sheik Omar Abdel Rahman, the spiritual leader of the radical group Jama'at al-Jihad, who was commonly known in the United States as "The Blind Sheikh." Rahman was indicted along with the Jama'at al-Jihad members who assassinated Sadat because he was accused of issuing a fatwa ordering Sadat's murder. Later, Rahman traveled to the United States, where he issued a fatwa that declared the robbing of banks and killing of Jews to be lawful in America. He called on Muslims to assail the West, "cut the transportation of their countries, tear it apart, destroy their economy, burn their companies, eliminate their interests, sink their ships, shoot down their planes, kill them on the sea, air, or land" (quoted in Wright 2007, 177).

Rahman, along with nine of his followers, was arrested on June 24, 1993, following the first World Trade Center bombing in February 1993. The Federal Bureau of Investigation (FBI) managed to record Rahman issuing a fatwa encouraging acts of violence against US civilian targets, particularly those in the New York and New Jersey metropolitan area. The most startling plan, the government charged, was to set off five bombs in 10 minutes, blowing up the United Nations, the Lincoln and Holland tunnels, the George Washington Bridge, and a federal building housing the FBI.

The importance of fatwas for promoting violence and terrorism is described by Kenan Malik (2009) in his book *From Fatwa to Jihad*. As Malik argues, the publication of both *The Satanic Verses* in 1989 and the Danish cartoons of the Prophet Muhammad in 2005 did not immediately lead to violent overreaction: it took quite a while, in each case, to carefully stoke the fires before a small number of opponents managed to fan the flames in just the right way to make for the conflagrations that followed. Ayatollah Khomeini's fatwa against Salman Rushdie was a turning and rallying point; the reactions to it, especially by European governments, suggested that such intemperate actions were a great way to get attention—and to get one's way. And, as Malik indicates, many have continued to play right into the hands of the small but vociferous extremist minority.

The Internet soon became a popular platform for Islamists to present their messages, including fatwas. There are numerous "online fatwa" websites, mostly in the form of answers given by Muslim authority. Online fatwas are particularly attractive to Muslims who do not have access to a qualified mufti. Naturally, this category especially applies to Muslims in the diaspora. Presumably, many of them look for a suitable fatwa website in the most common way: through the Internet. That means that high-ranking, visually attractive, well-advertised, or professional-looking pages are most likely to gain more visitors and thereby more influence. Some of these websites publicize their sites with slogans like "The most comprehensive online Islamic fatwa guide" and "1346 Fatwas Available" (FatwasIslam.com) or "Fatwa from the Major Scholars of the Muslim World" (fatwa-online.com). The numerous fatwas posted on these sites relate to daily Muslim practices, duties, and guidance. They are not violent, and they do not promote terrorism, suicide, or war. However, several jihadi fatwas started to emerge online, and today the Internet has become the most instrumental and effective instrument for spreading terrorist fatwas. In his article "The Internet Is the New Mosque," Abdallah el-Tahawy (2008, 10), a journalist at Islam Online, argues:

> Specifically, the Internet has become not only a clearinghouse for Koranic text, but also for religious guidance and even *fatwas* (religious edicts). This new, global online Islam has been propagated by countless websites maintained by sheikhs, religious scholars and even laymen. Today, any person can look up a *fatwa* on any subject, checking whether a particular action is *haram* (forbidden) or *halal* (permissible), sometimes within minutes, with just a few clicks of the mouse. Needless to say, this accessibility has been a boon to Islamic practice.

Jihadist Fatwas

The authors of jihadist fatwas come from diverse backgrounds. Some are scholars, some are religious authoritative figures, and others are political leaders of radical movements who are not seen in the wider Islamic world as having authority to provide fatwas, but are accepted as authorities by their own followers. Moreover, not all of the fatwas are initiated by individuals; some are published or posted online by Islamic institutions or by "fatwa committees" affiliated with certain Muslim communities or radical Jihadi groups.

Osama bin Laden issued two fatwas, one in 1996 and another in 1998. Both documents initially appeared in the Arabic-language London newspaper *Al-Quds Al-Arabi*. At the time, bin Laden was not a wanted man in any country except his native Saudi Arabia, and was not yet known as the leader of the international terrorist organization al-Qaeda. Therefore these fatwas received relatively little attention. Bin Laden's 1996 fatwa is entitled "Declaration of War against the Americans Occupying the Land of the Two Holy Places." It is a long declaration, documenting American activities in numerous countries (Ranstorp 1998). The 1998 fatwa was signed by five people, four of whom represented specific Islamist groups: Osama bin Laden and his deputy Ayman al-Zawahiri, and Ahmed Refai Taha and Abu Yasser (an alias) of the Egyptian Sunni Islamist group al-Gama'a al-Islamiyya. Mir

Hamzah, "Secretary of the Jamiat Ulema-e-Pakistan," and Fazul Rahman, "Emir of the Jihad Movement in Bangladesh" were also among the signatories. The signatories were identified as the "World Islamic Front for Jihad against Jews and Crusaders." This fatwa complains of American military presence in the Arabian Peninsula, and American support for Israel. It purports to provide religious authorization for indiscriminate killing of Americans and Jews everywhere. In this fatwa, faxed to the London newspaper *Al-Quds*, the group also wrote: "The ruling is to kill the Americans and their allies is an individual duty for every Muslim who can do it, in order to liberate the Al Aqsa mosque [Jerusalem] and the Holy Mosque [Mecca] . . . This is in accordance with the words of Almighty God. . . . We, with God's help, call on every Muslim who believes in God and wishes to be rewarded to comply with God's order to kill the Americans and plunder their money wherever and whenever they find it" (MideastWeb 2013).

The Internet became a useful platform for the posting of jihadi fatwas and interpretations of fatwas. As noted by Bar (2006b, 3), "The age of information has opened up a new venue for Muslims to acquire religious instruction without coming in direct contact with the consulting Sheikh. The Internet now allows a Muslim to send a query to any learned Sheikh by E-Mail and to receive his ruling either directly or in the public domain of websites dedicated to such fatwas." This trend was well documented by Gary Bunt's studies (Bunt 2000; 2003; 2009). According to Bunt, the Internet has profoundly shaped how Muslims perceive Islam and how Islamic societies today rely on online fatwas, social networking sites, blogs, and forums. Furthermore, the Internet has dramatically influenced forms of radical Islam and radical Islamic activism, including jihad-oriented campaigns and terrorist propaganda. Online terrorist fatwas have become instrumental platforms for such campaigns. Though most online fatwas are not related to terrorism, violence, radicalism, or jihad, terrorist groups have been using the Internet to post radical fatwas. There is a clear rise in the number of fatwas that declare jihad to be a religious obligation and define clear guidelines for waging jihad. Many of these online fatwas provide moral and religious justification for the use of terrorism and relate to terrorist issues, including the definition and identification of the battle space in which the attacks are to be executed, the identity of legitimate victims, the proper means of action, and the legitimacy of suicide attacks.

In November 2009, Quilliam, a think tank funded by the British Home Office, claimed that jailed jihadists in Britain are "strengthening jihadist movements" by posting online fatwas (BBC 2009). Abu Qatada, a radical Islamist cleric described by British domestic intelligence service MI5 as

"Osama Bin Laden's right-hand man in Europe," has published fatwas on the Internet from the maximum-security Long Lartin prison in Worcestershire, England, calling for holy war and the murder of moderate Muslims. Qatada, who is wanted on terrorism charges in Jordan, is held in the "supermax" segregation wing of the prison, which should be one of the most secure buildings in the country. Yet Qatada and Adel Abdel Bari, leader of the British branch of Egyptian Islamic Jihad, were able to smuggle out multiple fatwas legitimizing attacks by al-Qaeda and endorsing the murder of moderate Muslims and Muslims who are opposed to al-Qaeda.

The recent conflict in Syria has witnessed the emergence of radical Syrian Salafist groups—such as the al-Qaeda–linked Al-Nusra Front—and an influx of jihadists from Arab and European countries. Several jihadi clerics have ruled that the situation in Syria constitutes a defensive jihad, a war to repel a non-Muslim enemy that invaded a Muslim country (*jihad al-daf*). According to this fatwa, coming to Syria in order to wage jihad is a personal duty incumbent upon all able Muslim men (MEMRI Jihad & Terrorism Threat Monitor 2013a).[1] In one of the most powerful statements yet against Syrian president Bashar al-Assad and the Shiites, Islamic scholars issued the fatwa, declaring: "Jihad is necessary for the victory of our brothers in Syria—jihad with mind, money, weapons; all forms of jihad," adding: "What is happening to our brothers on Syrian soil, in terms of violence stemming from the Iranian regime, Hezbollah and its sectarian allies, counts as a declaration of war on Islam and the Muslim community in general" (quoted in Nasralla 2013). Earlier in 2013, a fatwa claimed to be issued by rebel groups in Syria allowed the fighters to rape non-Sunni women in Syria as part of a "sexual jihad." In March 2013, CNS News, a conservative US news outlet, reported that girls as young as 14 were being sent to Syria from other Middle Eastern and North African countries, following a fatwa issued by Saudi scholar Sheikh Mohamed al-Arifi for rebels to engage in "sexual jihad," a so-called "temporary marriage" that amounts to little more than sex slavery (Goodenough 2013). Though al-Arifi later backtracked after pressure, he had issued a fatwa saying that Syrian rebels can "temporarily marry" Syrian girls as young as 14, and promising "paradise" to the "wives" concerned. Later in 2013, another Islamic cleric publicly announced a fatwa that would permit jihadi rebels to rape women. Salafi Sheikh Yasir

---

[1] This ruling is similar to past rulings given by radical clerics with regard to various conflicts, most prominently by Palestinian Sunni scholar Abdullah Azzam during the war against the Soviet Union in Afghanistan in the 1980s.

al-Ajlawni's fatwa declared that it was legal for those Muslims fighting to topple secular Syrian president Bashar Assad and install sharia law to "capture and have sex with" all non-Sunni women, specifically naming Assad's own sect, the Alawites, as well as the Druze and several others—effectively, all non-Sunnis and non-Muslims (Ibrahim 2013). Although the reports of these "sexual jihad" fatwas come from conservative news sites, the online coverage indicates current concerns over the use of fatwas to sanction terrorist violence against both combatants and noncombatants.

It should be noted that terrorist fatwas are often meant to serve the terrorists' needs rather than to enforce sharia standards. An example of the flexible use of fatwas is the drug policy of the Taliban: When they were the uncontested rulers of Afghanistan in the 1990s, the Taliban outlawed drugs as entirely immoral. However, when facing urgent monetary needs to finance their ongoing insurgency against the NATO-led International Security Assistance Force coalition, the Taliban changed their "policy," using fatwas to support farming and drug trafficking in their dominions (Brahimi 2010, 9). This reconsideration proved to be quite effective, as the Taliban generated about $100 million per year from drug trading (Erwin 2008).

Cyber-Fatwas of Terror

As argued by numerous scholars, the role of radical online fatwas in legitimizing terrorism is a pivotal element in the social and political legitimization of terrorism and in the motivation of its supporters. Let us illustrate how these cyber-fatwas are related to justifying key issues in promoting terrorism: the use of suicide terrorism; the killing of innocents, women and children, or other Muslims; or the use of various weapons (including weapons of mass destruction and cyber-terrorism).

Who Are Legitimate Targets?

A common example of a legitimate target for jihad is the nonbeliever, a term that includes both Jews and Christians and often extends to foreigners in general. In March 2005, the Salafi Group for Call and Combat, today known as Al-Qaeda in the Islamic Maghreb (AQIM), issued a fatwa for jihad against foreigners in Algeria. The fatwa was signed by Abu Ibrahim Mustafa, the emir (prince) of the Salafi Group, who had taken over leadership of the group in October 2003 and had immediately pledged loyalty to bin Laden and al-Qaeda. According to the fatwa, "The Salafi Group states in these hard circumstances for the Muslim nation in general, and especially the mujahideen, to declare war on every foreigner nonbeliever in the Algerian lands. The governments in the Muslim lands are no more than flags put by the Crusaders before leaving to keep a watchful eye on the Muslims so they don't have a government which brings back the glory of Islam." The fatwa therefore calls for the killing of "the Jews and the Christians and all other nonbelievers" in Algeria, and calls on all Muslim Algerians to fight foreigners and disregard the local government: "Everyone, which concerns the individuals and establishments, is doing the duty for the victory of Islam and Muslims, is pushing away the attacks of Jews and Christians and other non-believers as they declare that they are not bound by any agreement with the converted Algerian government" (SITE Monitoring Service 2005). Abu Ibrahim Mustafa's fatwa was circulated on jihadist websites.

Jihadi fatwas also reinterpret the definition of civilians. According to al-Qaeda, in democracies the citizens are to blame for their government's decisions. A fatwa authorizing this view was issued by Sheikh Hammoud al-Uqla al-Shuyabi, the godfather of Saudi jihadis. As early as 2001, he published a fatwa declaring holy war against America and its supporters, and included members of the Saudi royal family among its targets. In his October 2001 fatwa, al-Shuyabi said that "whoever supports the infidel is considered an infidel" and "it is a duty to wage jihad on anyone who attacks Afghanistan." He answered a question posted online about when jihad, or holy war, is permissible: "Jihad is allowed against infidels like the Jews, Christians and atheists," he replied online (quoted in Tell 2006, 131). A similar fatwa came from Egyptian cleric Sheikh Yusuf al-Qaradawi, who has a history of activism in the Muslim Brethren. Forced from Egypt for his extreme views, he currently lives in Qatar. His fatwas are published and distributed primarily online. He issued numerous fatwas on jihad and violent activities, including one legitimizing the targeting of Israeli civilians. His ruling on Israelis included both men and women, since "an Israeli woman is not like women in our societies, because she is a soldier." Another example of a conveniently flexible definition of civilians concerns Iraq: Al-Qaradawi issued a fatwa permitting the abduction and killing of American civilians in Iraq in order to pressure the American army to withdraw from Iraq, arguing that "all of the Americans in Iraq are combatants, there is no difference between civilians and soldiers, and one should fight them, since the American civilians came to Iraq in order to serve the occupation. The abduction and killing of Americans in Iraq is a [religious] obligation so as to cause them to leave Iraq immediately" (MEMRI Jihad & Terrorism Studies Project 2004).

Online fatwas also provide rulings on the legitimacy of larger targets. A question posted in October 2010 on the Salafi-jihadi Minabr al-Tawhid wal-Jihad website, belonging to Salafi ideologist Abu Muhammad al-Maqdisi, asked about attacking Christians in Egypt: What is the ruling regarding attacking churches and blowing them up, and what is the ruling on attacking stores, vehicles, and other Christian property? What is the ruling with regard to intentionally or unintentionally attacking Christian women and children? The ruling, posted by Sheikh Abu al-Mundhir al-Shanqiti in November 2010, was that "if we narrow down the conflict, it is only permitted to attack those heading the attack against the Muslims, whether they are priests or not. But if we expand the conflict, then it is permitted to harm anyone in which there is an interest in attacking, according to the stages of the conflict and its severity. . . . This is because they [the Christians] are the ones who enabled the West to reach Islamic countries and served as a fifth column against Islam. Therefore, their blood is permitted as well as their money and property" (quoted in Jihadi Websites Monitoring Group 2010, 5–6). According to al-Shanqiti, Christians are the target in the conflict today because they are taking the position of the aggressor, not simply because they are Christian. Unlike al-Qaradawi's ruling on the status of Israeli women, al-Shanqiti states that children and women must not be harmed unless they took part in attacking Muslims. He explains that it is permitted if women and children are unintentionally harmed, but that one should be wise and prevent the enemy from using them as a shield. On December 31, 2010, al-Shanqiti posted a fatwa as a response to an individual who asked if Muslims are justified in attacking Christians in Muslim lands:

We clarify that these Christians present in Muslim lands are not the people of the covenant, and the rules of protection do not apply to them. Not every unbeliever born in a Muslim country or living there are of the people of the covenant, because protection under the covenant has description and conditions, which if not fulfilled, make a person outside the people of the covenant. We mention that the rulings of protection under the covenant do not apply to the unbelievers residing in lands of Islam today, because they do not pay the jizia [a tax imposed on non-Muslim subjects of an Islamic state] or abide by inferiority, and do not stop harming Muslims. Each one of these alone is sufficient to abrogate the covenant, because their blood was permissible from the start. As for the case of targeting them, it is up to the leaders of jihad in every country. If they decide to confront them in revenge for our Muslim sisters, then we should help them, and every Muslim who loves jihad must be an arrow in their quiver. (Quoted in Weimann 2011b, 771)

On April 12, 2013, al-Maqdisi's Minbar al-Tawhid wal-Jihad website posted a newer fatwa by Sheikh al-Shanqiti about the permissibility of bombing synagogues and churches. The fatwa was issued in response to a query by a reader who called himself "Assad al-Ma'arik," and asked specifically about bombing "the Jews' houses of worship in European states." Al-Shanqiti used the opportunity to address the general question of attacking Jewish and Christian houses of worship. The fatwa started by summarizing an opinion shared by many Islamic jurisprudents, namely that attacking houses of worship is illegitimate. This opinion was based on Qur'an 22:40: "Had Allah not defended some men by the might of others, the monasteries, churches, synagogues, and mosques in which His praise is daily celebrated would have been utterly destroyed." Al-Shanqiti rejected this view, arguing that the above verse refers only to the pre-Islamic era, as Judaism and Christianity both lost their validity with the appearance of Islam. However, although al-Shanqiti ruled it permissible to attack synagogues and churches in certain circumstances, he concluded that it is preferable to avoid this, for both religious and tactical reasons: "Attacking houses of worship, [even] in cases where it is legitimate, will be used as a pretext to defame jihad. Therefore, the mujahideen should not resort to this tactic except where there is an urgent need and necessity" (MEMRI Jihad & Terrorism Studies Project 2013a).

What Are Legitimate Actions?

Online fatwas allow terrorists to express their opinions on legitimate targets for terrorism. One such example is the case of Hamid Abdallah Ahmad al-Ali, a Kuwait-based terrorist recruiter who has provided financial support and ideological justification for al-Qaeda–affiliated groups seeking to commit acts of terrorism in Kuwait, Iraq, and elsewhere (United Nations Security Council 2009). On January 16, 2008, the United Nations Security Council listed al-Ali as being associated with al-Qaeda for "participating in the financing, planning, facilitating or perpetrating acts or activities by, in conjunction with, under the name of, or in support of . . . supplying, selling or transferring arms and related material to" and "recruiting for" al-Qaeda–affiliated cells in Kuwait. In his role as a recruiter for terrorist organizations, al-Ali has not only provided financial support for recruits (including paying their travel expenses to Iraq), but has also issued fatwas to justify acts of terrorism, including a fatwa endorsing suicide bombing operations and, more specifically, the flying of aircraft into targets during such operations. This fatwa sanctioned "the permissiveness, and sometimes necessity, of suicide operations on the conditions of crushing the enemy (or causing moral defeat to the enemy), to obtain victory." According to this fatwa, "In modern time(s) this can be accomplished through the modern means of bombing, or by bringing down an airplane on an important site that causes the enemy great casualties" (United Nations Security Council 2009).

Other online fatwas target iconic Western companies. In November 2010, Abu Muhammad al-Maqdisi's jihadi website Al-Minbar wal-Tawhid posted a fatwa permitting mujahideen to target companies owned by infidels, such as Coca-Cola and McDonald's. The fatwas on the Al-Minbar wal-Tawhid site came in response to a posting by a member named "Abu Sayyed Qutub," which included the following questions: What is the ruling regarding companies that distribute Jewish and American products, such as Coca-Cola and McDonald's? Is it permissible to receive help from gangsters in order to carry out jihad operations, as they say the jihad in Algeria is doing? And is a group of fewer than 10 young people without military or organizational experience permitted to target tourists in countries with apostate [Muslim] governments, where no known jihad group operates? Sheikh Abu Walid al-Maqdisi, a Gaza-based cleric, replied that according to Islamic law, it is forbidden to harm Muslim lives and possessions, but it is permissible "to target infidel and polytheist lives and possessions." "If the owners of these companies are Muslims, it is forbidden to harm or steal from them, even if they distribute or sell goods produced by the Jewish and Christian enemies of Allah, as long as the essence of the commerce and of the goods is permissible according to sharia," Abu Walid al-Maqdisi said. "However, if these companies are controlled by infidels, their property may be taken as booty, since the infidels of today are considered combatants" (MEMRI Jihad & Terrorism Threat Monitor 2010).

Abu Walid al-Maqdisi has provided rulings on other aspects of terrorist action. Kidnapping and killing tourists was permissible so long as it was done by "people who are reliable, knowledgeable in Sharia, and have organizational and military experience" and are acting for the benefit of Muslims, he said. If a Muslim lives in a place where no organized jihadist group exists, al-Maqdisi said, he should establish one. Finally, he added, Muslims are prohibited from obtaining help from infidel gangsters (although they may purchase weapons from them). Islamic law does not bar Muslims from receiving help from Muslim criminals, but according to al-Maqdisi, the assistance should be rejected "if it is likely to harm the reputation of the mujahideen, or their jihad plan" (MEMRI Jihad & Terrorism Threat Monitor 2010).

Online fatwas have also authorized cyberterrorism and attacks on websites. In October 2008, a fatwa published on the website of the Islamist Egyptian Muslim Brotherhood movement declared that top Muslim scholars have decreed that hacking and sabotage of American and Israeli websites are allowed under Islamic law and are a form of jihad or holy war. The fatwa was issued by a committee from the highest authority in Sunni İslam, Al-Azhar University in Cairo, Egypt, and was posted online as a response to numerous questions from radicals asking to be allowed to destroy Israeli and American state websites. "This is considered a type of lawful Jihad that helps Islam by paralyzing the information systems used by our enemies for their evil aims," the fatwa read. "This Jihad is not different from the armed one. In fact, it might be more important if you consider the global dimensions of the Internet. Whoever wins this war will become the strongest in the realm of information" (Adnkronos International 2008).

Finally, the media are a legitimate target for terrorism. An online fatwa issued in November 2013 by the militant Tehreek-e-Taliban Pakistan group approves death for television hosts, journalists, analysts, and media personalities whose acts and views are deemed by the Taliban to be against Islamic law. The fatwa was first authored in 2012 by a group of ulama (clerics) and muftis associated with the group, and reissued online (MEMRI Jihad & Terrorism Threat Monitor 2013b).

Are Suicide Operations Legitimate?

Most religions consider suicide to be an undesirable act. In Islam, the Qur'anic verse "Spend in the way of Allah; do not cast yourself into destruction" (Qur'an 2:195) is one example of a religious ruling against suicide. Nonetheless, jihadi fatwas have found creative ways to justify suicide bombing. In the "forbidding verse," the words "in the way of Allah" (*fi sabil Allah*, in Arabic) are interpreted as "for the sake of Allah." Thus, some fatwas claim that this means that suicide is acceptable if it is done for Allah. In these fatwas, the same verse that traditional Islam interprets as prohibiting suicide is interpreted as supporting suicidal actions if committed "in the way of Allah."

Another tactic in jihadist fatwas is to present suicide as a new conception of martyrdom. This shift challenges traditionally strong Islamic prohibitions against suicide (Qaradawi 2003). Bernard Freamon (2003, 300) argues that "this transformation of religious doctrine . . . resulted in the appearance of a new norm of jihadist battlefield behavior—self-annihilation—a norm that is now accepted as a valid discharge of religious obligation under the law of the military Jihad." Yusuf al-Qaradawi is a leading figure in online terrorist fatwas and a good example of the use of fatwas to legitimize terrorist actions. Here is a typical al-Qaradawi question-and-answer form of fatwa:

> The martyr operation is the greatest of all sorts of jihad in the cause of Allah. A martyr operation is carried out by a person who sacrifices himself, deeming his life [of] less value than striving in the cause of Allah, in the cause of restoring the land and preserving the dignity. To such a valorous attitude applies the following Qur'anic verse: "And of mankind is he who would sell himself, seeking the pleasure of Allah; and Allah hath compassion on (His) bondmen." (Qur'an, 2: 207). But a clear distinction has to be made here between martyrdom and suicide. Suicide is an act or instance of killing oneself intentionally out of despair, and finding no outlet except putting an end to one's life. On the other hand, martyrdom is a heroic act of choosing to suffer death in the cause of Allah, and that's why it's considered by most Muslim scholars as one of the greatest forms of jihad. When jihad becomes an individual duty, as when the enemy seizes the Muslim territory, a woman becomes entitled to take part in it alongside men. Jurists maintained that when the enemy assaults a given Muslim territory, it becomes incumbent upon all its residents to fight against them to the extent that a woman should go out even without the consent of her husband, a son can go too without the permission of his parent, a slave without the approval of his master, and the employee without the leave of his employer. This is a case where obedience should not be given to anyone in something that involves disobedience to Allah, according to a famous juristic rule. . . . To conclude, I think the committed Muslim women in Palestine have the right to participate and have their own role in jihad and to attain martyrdom. (Middle East Forum 2004)

Al-Qaradawi's fatwa on suicide was echoed in numerous other terrorist fatwas. For example, Abd al-Aziz al-Rantisi, then one of the leaders of Hamas, argued that "suicide depends on volition. If the martyr intends to kill himself, because he is tired of life—it is suicide. However, if he wants to sacrifice his soul in order to strike the enemy and to be rewarded by Allah—he is considered a martyr (*shahid*). We have no doubt that those carrying out these operations are martyrs" (quoted in MEMRI Jihad & Terrorism Studies Project 2001). Al-Rantisi based this distinction on al-Qaradawi's fatwa.

Counter-fatwas have challenged the legitimization of suicide attacks. In August 2005, the Syrian cleric Abu-Basir al-Tartusi posted a fatwa online under the headline "A Word of Warning About Suicide Operations." He argued: "I have received 1,000 questions about these operations, which are for me closer to suicide than martyrdom. They are haram (forbidden) and impermissible, for several reasons." Al-Tartusi, who lives in London, cited in the fatwa some of the Prophet Muhammad's sayings, among them: "Anyone who harms a believer has no jihad" (quoted in Bunt 2009, 219). The fundamentalists launched a bitter attack on al-Tartusi on their websites and accused him of letting down al-Qaeda's supporters. One of them asked, "What do you expect from him when he lives in London?" Another posting argued, "One should not get attached to these people because they did not fight before. The rules on jihad are taken from the mujahideen. I never thought of learning about jihad from those sitting who are used to issuing fatwas from London."

Is the Use of Weapons of Mass Destruction Legitimate?

Online fatwas have also been dedicated to considering terrorist use of weapons of mass destruction (WMD). In May 2003, the young Saudi cleric Shaykh Nasir bin Hamid al-Fahd published "A Treatise on the Legal Status of Using Weapons of Mass Destruction Against Infidels" (Fahd 2003). Al-Qaeda and its supporters have used this fatwa as a justification and authorization for using weapons of mass destruction against infidels—in this case, against the United States. Al-Fahd begins by describing the term "weapons of mass destruction" as an "inexact term," claiming that chemical, biological, or nuclear weapons that killed a thousand people would be called "internationally banned weapons" by the West, whereas the use of "high explosive bombs weighing seven tons apiece and [that] killed three thousand or more" would be called "internationally permissible weapons." On that basis, he dismisses the West's treaties and regulations banning WMD proliferation as mere attempts to scare others and protect itself. "Thus it is evident," he wrote, "that [the Western nations] do not wish to protect humanity by these terms, as they assert; rather, they want to protect themselves and monopolize such weapons on the pretext of banning them internationally." As a result, he argues, "all these terms have no standing in Islamic law, because God

Almighty has reserved judgment and legislation to Himself. . . . This is a matter so obvious to Muslims that it needs no demonstration. . . . In judging these weapons one looks only to the Koran, the Sunnah [the sayings and traditions of the Prophet], and the statements of Muslim scholars." Al-Fahd also argues that large civilian casualties are acceptable if they result from an attack meant to defeat an enemy, and not an attack aimed only at killing the innocent: "The situation in this regard is that if those engaged in jihad establish that the evil of the infidels can be repelled only by attacking them at night with weapons of mass destruction, they may be used even if they annihilate the infidels" (Fahd 2003).

In 2008, Ayman al-Zawahiri released his book *The Exoneration*. In it, he resurrected the WMD fatwa issued by Nasir al-Fahd in 2003. Al-Zawahiri adopted al-Fahd's ideas to reach the same conclusion: The use of nuclear weapons would be justified as an act of equal retaliation, "repaying like for like." Al-Zawahiri raised key Qur'anic themes to sweep away all potential objections to the use of WMD and adopted al-Fahd's examples word for word. The Prophet Muhammad's attack on the village of al-Taif using a catapult, for instance, permits the use of weapons of "general destruction" that are incapable of distinguishing between civilians and combatants (Mowatt-Larssen 2010). By echoing the previous fatwa on the subject, al-Zawahiri's fatwa reinforces the legitimacy of the use of WMD in global jihad.

Radical Islam has adopted the Internet as a favorite platform for spreading jihadi fatwas, legitimizing the use of violence, suicide attacks, targeting innocents, and even of killing moderate Muslims. By learning the flow of fatwas and their sources and reasoning, and by monitoring the debates over fatwas, one can find practical ways to combat the misleading and illegitimate uses to which they have been put.

Chapter 7

Terror on Social Media

On the evening of March 1, 2011, Arid Uka, an Albanian Muslim living in Germany, was watching videos on YouTube. Like many before him, he watched a video clip that presented the gruesome rape of a Muslim woman by US soldiers—a clip edited and posted on YouTube for jihadi propaganda purposes. Within hours of watching the video, Uka took a handgun and boarded a bus at Frankfurt Airport, where he killed two US servicemen and wounded two others.

Uka's Internet activity—most obviously, his Facebook profile—revealed a growing interest in jihadist content. His increasing self-radicalization led up to the point where, after watching the aforementioned video, he decided to take action in an alleged war in defense of Muslims. Ulka was not a member of a terrorist organization, nor did he visit one of the infamous training camps for terrorists. His entire process of self-radicalization, from his

---

This chapter expands on material originally published in "Terror on Facebook, Twitter, and You-Tube" (Weimann 2010a) and *New Terrorism and New Media* (Weimann 2014a).

early attraction to jihadi preaching to his preparations for the final deadly mission, was performed online (Weimann 2014a).

Arid Uka is a typical case of new media creating new terrorists. Internet-savvy terrorists have learned to use the newest online platforms, commonly known as "new media" or "social media," for their efforts. This chapter examines interactive online communication on social media venues and how it is used by terrorists and their supporters, with a particular focus on the leading platforms of Facebook, Twitter, and YouTube. The analysis is based on the database collected during a 15-year project of monitoring thousands of terrorist websites and online forums, chatrooms, and social networking sites. Given the rather broad presence and use of new online media by terrorists, it will present only illustrative examples, highlighting the most recent developments.

What Is Social Media?

Social media refers to the interaction among people in which they create, share, and/or exchange information and ideas in virtual communities and networks (Ahlqvist et al. 2008). Social media depends on new communication technologies such as mobile and web-based networks to create highly interactive platforms through which individuals and communities share, co-create, discuss, and modify user-generated content. As a result, social media are immersive. When we are on a social media site, we feel that we are virtually together with our friends, relatives, and colleagues. Online interaction brings people closer, faster, by allowing users to contact each other rapidly online and then, when desired, move offline. In the case of two people, each seeking a companion, the result may be a happy union. In the case of aspiring terrorists, the result may be less positive.

Social media differs from traditional or conventional media in many aspects such as interactivity, reach, frequency, usability, immediacy, and permanence (Morgan, Jones, and Hodges 2012). Unlike the "one-to-many" communication of traditional media such as newspapers, television, and film—where the audience might be virtually limitless but where a small cohort of established institutions selectively disseminates information—social media enables anyone to publish or access information. With social media, information consumers also act as communicators, vastly expanding the number of information transmitters in the media landscape (Amble

2012). This comparatively inexpensive and accessible process has lowered the barriers to enter communication markets by letting in small, diffused sets of communicators and groups, a form of distribution known as the "long tail" (Anderson 2006).

The growing use of social media is impressive. In the United States, the total time spent on social media increased by 37 percent, from 88 billion minutes in July 2011 to 121 billion minutes in July 2012 (Nielsen 2012). Popular social media tools and platforms include the following:

- Blogs: A platform for casual dialogue and discussions on a specific topic or opinion.
- Facebook: The world's largest social network, with more than 1.3 billion users as of June 2014 (Facebook 2014). Users create a personal profile, add other users as friends, and exchange messages, including status updates. Brands create pages and Facebook users can "like" brand pages.
- Twitter: A social networking/micro-blogging platform that allows groups and individuals to stay connected through the exchange of short status messages, with a 140-character limit.
- YouTube and Vimeo: Video hosting and watching websites. The 2012 Noel-Levitz *E-Expectations Report* found that 62 percent of high school juniors and seniors watch YouTube videos at least once a week (Noel-Levitz 2012).
- Flickr: An image and video hosting website and online community. Photos posted on Flickr can be shared on Facebook, Twitter, and other social networking sites.
- Instagram: A free photo and video sharing app that allows users to apply digital filters, frames, and special effects to their photos and then share them on a variety of social networking sites.
- LinkedIn Groups: A professional social networking website where groups of professionals with similar areas of interest can share and participate in conversations about things happening in their fields.

Social media is often associated with benefits such as creating new social connections and improving communication, yet such positive outcomes are not always the case. An increase in social media offers and usage has been correlated with an increase in negative, abusive practices and outcomes such as cyberbullying, online sexual predators and sexual abuse, and more. Terrorism is certainly one of the darkest sides of the new social media.

When New Terrorism Met New Media

Before the existence of the Internet, terrorists' social networking was based on conventional human interaction that took place in key locations such as schools, marketplaces, religious centers, and houses of worship. Consequently, for traditional terrorist organizations like the first generation of al-Qaeda, critical social networking hubs consisted of secret meeting places, guesthouses, extremist mosques, and fixed training camps. In the wake of the September 11, 2001, terrorist attacks on the United States, these traditional hubs were quickly targeted, and in the face of constant pressure from counterterrorism agencies, al-Qaeda, its affiliates, and other terrorist organizations have moved to cyberspace and to new online platforms.

Terrorists have three major reasons to use these platforms. First, these platforms are by far the most popular venues for a mainstream audience in an increasingly digital age. As the most important purposes of terrorist Internet use are propaganda, radicalization, and recruitment, it makes sense to follow the mainstream audience. Second, social media companies provide a user-friendly and reliable service for free. Hosting a separate personal or organization website requires time, money, and a slightly higher level of Internet expertise. In addition, once online, terrorist websites tend to come under attack by law enforcement agencies or civil activists, making it more difficult for the operators to maintain a stable service. By using a social media platform as a host, terrorists can mitigate some of these effects. Finally, whereas the older versions of terrorist websites effectively were waiting for visitors to arrive, a social networking approach allows terrorists to reach their target audiences and virtually "knock on their doors" (Weimann 2014a).

Social networking online is becoming more attractive for terrorists who are targeting potential members and followers. These types of virtual communities are growing increasingly popular all over the world, especially among younger demographics. Social networking sites allow terrorists to disseminate propaganda to an impressionable age bracket that might empathize with their cause and possibly agree to join. Jihadist terrorist groups especially target youths for propaganda, incitement, and recruitment purposes. Increasingly, terrorist groups and their sympathizers are using predominately youth-dominated, Western online communities like Facebook, MySpace, and Second Life and their Arabic equivalents to reach out to the younger generation. Counterterrorism expert Anthony Bergin says that terrorists use these websites as recruitment tools "in the same way a pedophile might look at those sites to potentially groom would-be victims" (quoted in *The Advertiser* 2008).

Many social media users join interest groups, and these groups enable terrorists to target users whom they might be able to manipulate. These users often accept people as "friends" on the social media site whether or not they know them, thereby giving strangers access to personal information and photos. Some people even communicate with the strangers and establish virtual friendships. Terrorists can therefore apply the narrowcasting strategy (see chapter 2) used on the broader Internet to social networking sites. They can tailor their name, accompanying default image, and information on a group message board to fit the profile of a particular social group. Interest groups also provide terrorists with a list of predisposed recruits or sympathizers. In the same way that marketing groups can view a member's information to decide which products to target on their webpages, terrorist groups can view people's profiles to decide who they are going to target and how they should configure the message.

It is evident that as social networking has quickly become a ubiquitous part of many people's lives worldwide, it too has enjoyed increased significance as an amplifier for violent extremist viewpoints, and as a way for terrorists and their supporters to identify each other and build new relationships (Kohlmann 2011). In March 2010, one user on al-Qaeda's Fallujah Islamic Network online forum declared, "The least we can do to support the Mujahideen is to distribute their statements and releases." He added, "We wish from the brothers to also distribute the statement via YouTube and widely . . . and on Facebook." The user offered a cautionary note about using Facebook: "The suggested method is to always access it via proxy, otherwise you're in danger. Make one email on Yahoo that's dedicated for the [online] battle only. After creating the email, register on Facebook under a pseudonym with the email you created, and via which the account will be activated. Search for all the profiles and groups."[1]

In their closed forums, jihadists offer sophisticated reflections on how to strategically exploit the advantages and avoid the disadvantages of the new media for the sake of their cause. On January 4, 2012, on the leading jihadist forums al-Fida and Shumukh al-Islam, a comprehensive paper on "Electronic Jihad" was published. The paper provided an analysis as well

---

[1] Fallujah Islamic Network, http://www.al-faloja.info/vb/showthread.php?t=105942 (accessed March 10, 2014).

as an overview of the goals and means of using the Internet. In a remarkable conclusion, the author stated:

> [. . .] any Muslim who intends to do jihad against the enemy electronically, is considered in one way or another a mujahid, as long as he meets the conditions of jihad such as the sincere intention and the goal of serving Islam and defending it, even if he is far away from the battlefield. He is thus participating in jihad indirectly as long as the current contexts require such jihadi participation that has effective impact on the enemy (SITE Monitoring Service 2012b).

Considering that the divine status of a "mujahid" is presumably one of the key attractions for young people to participate in terrorist actions, this conclusion is striking. For militant Islamists, the threshold to give up a familiar, comfortable life to travel to an actual battle zone and risk injury or death is understandably high and keeps many from doing so. However, the threshold for engaging in "electronic jihad" is far lower. If the attitude prevails that, in the eyes of God and the people, online activism is a proper, respectable, and sufficient form of jihad, one can expect further increasing efforts in online propaganda (and cyberattacks)—which, in turn, may lead to even more radicalized individuals and ultimately new attacks. This type of call for electronic jihad has not gone unanswered: in July 2013, the Al-Battar Media Battalion was established, and claimed to be exclusively devoted to disseminate propaganda in "the service of jihad and the mujahideen" (SITE Monitoring Service 2013d).

The emergence of special groups dedicated to online activism but not affiliated with any particular group, such as the Al-Battar Media Battalion and the Nukhbat Al-I'lam Al-Jihadi (Jihadi Media Elites), marks a new waypoint in jihadists' professional use of the new media. Individuals such as Jose Pimentel or the user known as "Jihad Princess," who apparently never actually engaged in jihadist violence (although Pimentel tried; see chapter 3) but nevertheless enjoyed celebrity-like status online because of the manifold, high-quality material they posted, can similarly be regarded as indicators of this trend (SITE Monitoring Service 2011c). Yet terrorists are well aware of the risks involved in this online activism. A member of an English-language jihadi forum issued a warning, reminding readers that a Facebook network would allow security agencies to trace entire groups of jihadists:

> Don't make a network in Facebook . . . Then Kuffar will know every friend you have or had in the past. They will know location, how you

look, what you like, they will know everything! Join Facebook if you want and use it to keep in touch with friends and brothers far away but not as a network. (SITE Monitoring Service 2009)

A more thorough assessment of this approach to online terrorism requires more in-depth assessment of terrorist uses of three specific social media platforms: Facebook, Twitter, and YouTube.

Facebook Terrorism

Facebook is a free online social networking service. After registering, users can compile their personal profile out of standardized data, self-composed texts, pictures, and videos. Facebook offers convenient functions to interact with other users, such as becoming "friends" with them, which, depending on the users' setting, may allow access to additional private material on one's profile. Using Facebook, it is easy to comment on other users' content, or to open up forums or groups on specific topics.

Membership within the international Facebook community has boomed in recent years. Facebook is currently the world's most popular social networking website, with an estimated 1.11 billion users worldwide, of whom the majority (54 percent) frequent Facebook on a regular basis. The average age of a Facebook user is about 30 years, which includes a 66 percent membership increase within the Middle East and a 23 percent increase in Asia (Associated Press 2013). Terrorists have taken note of the trend and have set up Facebook profiles as well. The main motivation for terrorists to use Facebook has been properly outlined by the terrorists themselves:

This [Facebook] is a great idea, and better than the forums. Instead of waiting for people to [come to you so you can] inform them, you go to them and teach them! . . . [I] mean, if you have a group of 5,000 people, with the press of a button you [can] send them a standardized message. I entreat you, by God, to begin registering for Facebook as soon as you [finish] reading this post. Familiarize yourselves with it. This post is a seed and a beginning, to be followed by serious efforts to optimize our Facebook usage. Let's start distributing Islamic jihadi publications, posts, articles, and pictures. Let's anticipate a reward from the Lord of the Heavens, dedicate our purpose to God, and help our colleagues. (Quoted in Department of Homeland Security 2010)

Noting the prevalent use of Facebook globally, on December 9, 2008, a jihadist posted a call for "Facebook Invasion" on the password-protected jihadist Al-Faloja forum. The posting notes the "great success" thus far in publishing jihadist media on YouTube, and urges jihadists to maintain that campaign on Facebook as well. Through the "Facebook Invasion," the jihadist hopes to reach Muslims and Americans, and "fight the media offensive on jihadist media, its forums, and its websites." The posting also included images detailing the registration process, adding friends, and setting up groups. The "Facebook Invasion" has argued for it as "a good way to first, reach the misled American people, and second, to reach the vast people's base among Muslims. It may be a new technique and a new field that we did not wage before, or for the security of the site and the arrest of many among the organizers of strikes. If it is a new technique, we will use it and master it, with permission from Allah" (quoted in Kennedy and Weimann 2011).

Although the often-conspiratorial jihadist forums are frequented mainly by hardcore jihadist sympathizers, the mainstream Islamic youth—whose radicalization is a key concern of jihadist terrorists—is on Facebook. Accordingly, a primary purpose of Facebook is to link occasional viewers of jihadist contents to the more hardcore closed forums. This is most obvious in cases where jihadist forums maintain Facebook pages under their own names (Department of Homeland Security 2010). A prime example is the Ansar al-Mujahideen forum, which mirrors all of its content to a Facebook page named "Ansar al-Mujahideen Network" via a RSS (Really Simple Syndication) feed program. It is widely believed that for the recruitment of terrorists, social ties are more important than genuine ideological convictions. This structure coincides with the underlying design of Facebook, where a high number of friends signifies a user's high status, and therefore friend requests are often accepted indiscriminately. Terrorists thereby have the opportunity to actively approach and recruit suitable users. Facebook's function as a gateway is well illustrated by the self-contradictory profiles of nascent terrorists that depict their split between the normal and jihadist worlds. In one instance, a Swedish suicide bomber "liked" the expressions "the Islamic Caliphate State," "the Islamic Day of Judgement," and "I love my Apple iPad" (Amble 2012).

Generally, two types of Facebook profiles with terrorist content can be identified: official and unofficial profiles. "Official" pages are often introduced with a statement issued by the respective group in the usual forums and media. An example is the "Al-Thabaat" page, which appeared on May 5, 2013, and in its "About" section straightforwardly described itself as

"Jihadi page for the group, Ansar al-Islam" (SITE Monitoring Service 2013c). Unsurprisingly, the page offers links to the official Ansar al-Islam forum and Twitter account. "Unofficial" profiles, by contrast, are mostly maintained by sympathizers who disseminate propaganda or instruction material. However, these breeds of profiles can be interrelated. An example of this interrelation is the Facebook page of Yemeni journalist Abd al-Razzaq al-Jamal, who repeatedly had posted on his page exclusive al-Qaeda material by others, or posted it himself. Included in al-Jamal's postings were excerpts of the upcoming issue of al-Qaeda's online *Inspire* magazine and a letter claiming full responsibility for an attack in Yemen (MEMRI Jihad & Terrorism Studies Project 2013b).

According to a 2010 special report by the US Department of Homeland Security, terrorists use Facebook for several purposes:

- As a way to share operational and tactical information, such as bomb recipes, weapon maintenance and use, tactical shooting, etc.
- As a gateway to extremist sites and other online radical content by linking on Facebook group pages and in discussion forums.
- As a media outlet for terrorist propaganda and extremist ideological messages.
- As a wealth of information for remote reconnaissance for targeting purposes. (Department of Homeland Security 2010)

A wide range of instruction material is disseminated on Facebook, including tactical shooting and handling instructions for AK-47 assault rifles, manuals for manufacturing improvised explosive devices, and several chemical recipes for poisonous substances. This material was disseminated through Facebook groups affiliated with jihadist content. Additionally, to a certain degree terrorists will use Facebook for operative purposes to both coordinate and advertise their actions. For instance, in May 2013 the Tunisian group Ansar al-Sharia posted a map with close-to-real-time information about police checkpoints, and possible routes to avoid them, in order to direct its followers to an illegal conference. The Afghan Cyber Army is another example of a group that makes particular operational use of Facebook. The group, whose motto (or one of whose mottos) is "Live for Afghanistan or leave Afghanistan," actively maintains what it claims are official accounts on Facebook and YouTube. In its Facebook "About" section, the group says that it was formed in January 2005; that it is located in Kabul City, Afghanistan; and that it aims to "secure Afghan Cyber Space from cyber criminals and Afghanistan enemies." During 2013, the Afghan Cyber Army hacked the official US government portal usa.gov, US bank and credit card accounts and more than a million US Facebook accounts, and other US websites. It also has posted its claims of successful hacks on its Facebook page, sometimes accompanied by reports from various media confirming its claims. On November 15, 2013, the group claimed on its Facebook page that it had hacked a number of Pakistani websites related to airlines, insurance companies, e-commerce, real estate, and health, and warned that it would be hacking many thousands more (Stalinsky and Sosnow 2013a).

In addition to these specific examples of terrorist action, it is feared that terrorists could use Facebook for remote target reconnaissance. Terrorists can use social networking sites such as Facebook to monitor military personnel. Many soldiers unwittingly post detailed information about themselves, their careers, family members, date of birth, present locations, and photos of colleagues and weaponry. Even if the information does not give details about the logistics of troop movements, it could potentially endanger the friends and relatives of the military and security personnel. In 2008, in an unprecedented New Year "high priority" warning, the British domestic security service MI5 asked British troops to remove all personal details they had posted over the Christmas period on the Facebook, MySpace, and Friends Reunited social networking sites. MI5's Internet analysts had discovered al-Qaeda operatives had been monitoring the sites to gather details that could be used to launch terror attacks (Sammy67 2008). Both US and Canadian troops have been asked to exclude any information from their profiles that might link them with the military, and have been warned about posting certain details or photos on their social networking profile pages. Israeli intelligence also determined that Hezbollah had been monitoring Facebook to find potentially sensitive information about Israeli military movements and intelligence that could harm Israeli national security. An Israeli intelligence official admitted that "Facebook is a major resource for terrorists, seeking to gather information on soldiers and IDF [Israel Defense Forces] units and the fear is soldiers might even unknowingly arrange to meet an Internet companion who in reality is a terrorist" (*Middle East Times* 2008). In fact, the fate of the Israeli teenager Ofir Rahum offers an example of how social media can be used to conduct an actual terrorist plot. In January 2001, following a number of online exchanges with a woman who claimed to want to meet him for romantic purposes, Rahum was lured to enter the West Bank, where Palestininan terrorists were waiting to murder him (Hershman 2001).

Maybe more than any other jihadist group, Hezbollah has learned to capitalize on Facebook's accessibility and global usage. As early as 2008, Hezbollah had established a presence on Facebook, which has since grown into hundreds of pages, available in no fewer than nine languages and with tens of thousands of fans. This Facebook network of Hezbollah pages includes groups dedicated to Hezbollah itself, as well as to its leaders (such as Secretary General Hassan Nasrallah and assassinated security head Imad Mughniyeh), and helps to spread Hezbollah propaganda around the world (SITE Monitoring Service 2014c). Hezbollah's international reach on Facebook is evident in pages like the Indonesian-language Hezbollah page founded on March 29, 2010, which has attracted hundreds of fans; the Turkish-language "Hizbullah" founded on November 10, 2010, which already has more than 140 fans; and many more Hezbollah Facebook pages in French, Hebrew, Albanian, Spanish, Urdu, and Persian. The vast majority of Hezbollah pages, however, are in Arabic or English. In these pages, Hezbollah's versatility is demonstrated, as the group and its supporters have created pages not only for the organization itself, but also for its leaders. Hezbollah secretary general Nasrallah has inspired dozens of groups and profiles, some with members or friends exceeding 1,700. According to a 2011 SITE Monitoring Service report, updated in 2014, "In all, Hezbollah's Facebook network has perhaps 100,000 fans spread out across several hundred pages" (SITE Monitoring Service 2014c, 3) The leading Facebook pages are "We Are Hizbullah," "Hezbollah," and "Hezbollah will not be stopped," which serve as archetypes for Hezbollah Facebook pages in general. "We Are Hizbullah" focuses on official propaganda from Al-Manar, Hezbollah's media wing, while "Hezbollah will not be stopped" glorifies martyrdom and prominent leaders of Hezbollah. "Hezbollah" is a mix of the two, providing both official media and incitements to violence.

Since 2011, the ongoing uprising in Syria has involved several groups fighting against Syrian president Bashar al-Assad. Some of these groups are related to terrorist groups such as Hezbollah, and some are involved in the global jihadi movement, including a group representing al-Qaeda. These groups are turning to social media such as Facebook for propaganda, psychological warfare, and tutorials on how to use their weapons. Jabhat al-Nusra (Al-Nusra Front), a branch of al-Qaeda operating in Syria, has been designated a terrorist organization by Australia, the United Nations, the United States, and the United Kingdom (Agence France-Presse 2013). The head of the Al-Nusra Front, Abu Mohammed al-Golani, extensively uses Facebook and other online social media. For example, in August 2013 he vowed unrestrained rocket attacks on Alawite communities, alongside attacks on President Assad's government in revenge for an alleged chemical weapons strike launched by Assad's army. Al-Golani's message was posted on Facebook, Twitter, and on the militant websites that usually broadcast the views of al-Qaeda and similar extremist groups. Al-Nusra Front had its own Facebook page (facebook.com/jalnosra), which contained releases, photographs, and videos from the fighting in Syria, eulogies for the organization's martyrs, news on the fighting on the ground, and so forth. The apparently most recent trend on Facebook is virtual eulogies for killed ("martyred") jihadists fighting in Syria, which are often also posted on Flickr (Stalinsky 2013c). These eulogies present the fighters as role models for Muslims and posthumously make them immortal—an appealing prospect for radical Muslims who may be marginalized in their current societies (Jost and Covi 2013).

There is evidence of a growing sophistication in terrorist use of Facebook. On the Shamouhk al-Islam forum on June 2, 2011, a jihadist advised his companions how to attract a more mainstream audience by naming the account in a misleadingly innocent way:

> So, for example, the page "I am a Muslim, Praise be to Allah" has 24,522 fans, even though it doesn't spread valuable materials amongst Muslims. Meanwhile, "The Global Network of Jihad," doesn't exceed 500 fans despite the importance of what it publishes. (SITE Monitoring Service 2011a)

Jihadist Facebook pages show such examples of creative naming. For instance, two pages modified the world-famous Egyptian civil rights slogan "We are all Khaled Saeed"[2] to "We are all Usama bin Laden" and "We are all Jabhat al-Nusra." Furthermore, some online jihadists have even demanded not to maintain a low profile, but to do quite the opposite: They called on users to post the upcoming *Inspire* magazine on the Facebook pages of several public US figures or institutions, such as First Lady Michelle Obama or the US Army. They even offered respective link lists and online tutorials. The use of Facebook involves some risk from the terrorists' perspective.

---

[2] Egyptian blogger Khaled Saeed was arrested and beaten to death by police officers in Alexandria on June 6, 2010. The Facebook page "We are all Khaled Saeed," established in his memory, gained hundreds of thousands of followers and was a rallying point for Egyptian dissidents, contributing to the 2011–12 Egyptian revolution.

In jihadist forums, users outline relatively sophisticated measures to avoid detection, such as using entirely faked personal data and running anonymization software when browsing. Some even advise against posting jihadist pictures at all (Weimann 2014a, 7).

Facebook's terms of use prohibit any kind of criminal, violent, or hate content. However, it allows violent images if used for genuine information or "in a manner intended for its users to "condemn" the acts rather than celebrate them" (Oreskovic 2013). First, to enforce this prohibition—however comprehensible—standard, a case-by-case assessment is required, which is simply unreal given the huge overall number of posts on Facebook each day. Second, jihadists often post images of Muslim victims of Western warfare in Afghanistan, Gaza, or Iraq that are indeed meant to condemn the acts but are presented in a context that rather advocates violent "resistance," including terrorism. In spite of these difficulties, Facebook makes an effort to detect and delete hate content. A respective evaluation conducted by the Simon Wiesenthal Center grades the company's progress with an "A-" (Pollack 2013). It is notable that hate content spreads much more slowly in the (easier to monitor) English and Spanish sections of Facebook than in the Arabic-language one (Department of Homeland Security 2010). Jihadists do feel the impact of this control: After several of its posts were deleted, the jihadist Islamic News Agency expressed its outrage by "threatening" not to use Facebook any longer. The operators of the page "Third Palestinian Intifada" even felt the need to preventively threaten, rather ambitiously: "If Facebook Blocked This Page . . . All Muslims Will Boycott Facebook For Ever." (SITE Monitoring Service 2014d).

Twitter Terrorism

Twitter is a free microblogging service that allows every account holder to distribute messages which are limited in length to 140 characters. These "tweets" can be entirely open to the public or addressed only to those who express their particular interest in an author by "following" him or her (subscribing to his or her tweets). Twitter can be accessed via its website or respective applications for mobile devices. By mid-2013, the approximately 550 million Twitter users tweeted an average total of 9,100 messages every second. Twitter users post tweets at a rate of 250 million per day, and this number is growing rapidly (Statistic Brain 2013). Different from Facebook, Twitter is especially suitable for momentary, occasional users, who account for 72 percent of its users. Forty-three percent of Twitter users are between 18 and 34 years old (PearAnalytics 2009).

Recently, Twitter has emerged as the Internet application most preferred by terrorists, even more popular than self-designed websites or Facebook (Khayat 2013). Twitter is used mostly for propaganda, and to a lesser extent for internal communication. Terrorist use of Twitter has coincided with a recent trend in news coverage that often sacrifices validation and in-depth analysis for the sake of near-real-time coverage. As a result, mainstream media tend to use tweets as a legitimate source, especially when there is a lack of more valid or confirmed sources. Terrorists have repeatedly and methodically exploited this shortcoming for propaganda purposes. A prime example of this approach is the Syrian Electronic Army's hacking of the Associated Press Twitter account on April 23, 2013. The hackers tweeted the fake breaking news of a bomb attack inside the White House that injured Barack Obama. Immediately after this hoax, Wall Street suffered $136 billion in losses due to panicked investor reactions (Foster 2013).

Twitter can also provide a platform for terrorists to announce their intentions and actions to a broad, international audience. In September 2013, the Somali al-Qaeda–affiliated terrorist group al-Shabaab killed 72 people in a terrorist attack on the Westgate shopping mall in Nairobi, Kenya—and, in an historic first, the group used Twitter to not only claim responsibility for it but also give a live commentary on their actions. Several hours into the assault, the following tweet was posted on an al-Shabaab Twitter account: "The Mujahideen ('holy warriors') entered Westgate mall today at around noon and they are still inside the mall, fighting the Kenyan kuffar ('infidels') inside their own turf." It was the first confirmation that the attack was the work of al-Shabaab, and journalists around the world quickly reported this information. The group then quickly tweeted its rationale for the attack and gave operational details of the assault—all in real time. According to the London-based International Centre for the Study of Radicalization and Political Violence, which has studied al-Shabaab's use of social media, the group has been active on Twitter since December 2011, sending out a steady stream of tweets to at one point more than 15,000 followers, including a good number of journalists and terrorism analysts (Bergen 2013).

However, terrorists' main use for Twitter is to communicate with sympathizers. As Nico Prucha and Ali Fisher (2013) found in their analysis, Twitter has become the main hub for the active dissemination of links directing users to digital content hosted on a range of other platforms. An examination of 76,000 tweets produced by the al-Qaeda–related Al-Nusra Front in

Syria revealed that the tweets contained more than 34,000 links, many of which led to other jihadist digital contents (Prucha and Fisher 2013). Al-Nusra's official Twitter account, @jbhatalnusra, enjoyed a steady increase in the number of followers. Only one day after its creation on August 25, 2012, the account had more than 24,000 followers. Al-Nusra's tweets are updates from the various theaters of operations as well as propaganda releases and links to jihadi sites. In a similar vein, the eleventh issue of al-Qaeda's online *Inspire* magazine was publicized on Twitter, after the common practice of uploading the magazine to jihadi forums had become more and more difficult. In another instance, the group Minbar at-Tawhid wal-Jihad extensively used applications such as Twishort and TwitMail to link viewers to additional material, such as comprehensive fatwas. Through this sophisticated use, followers can read legal opinions on topics such as "Leaving one's country to wage jihad in another country's battlefield" or "Rulings on using stolen money for jihad" (Stalinsky 2012a). On April 18, 2013, Al-Qaeda in the Islamic Maghreb (AQIM) even hosted an online press conference over Twitter, in which participants could post questions that AQIM then answered in a PDF document (published again via Twitter) one week later (Prucha and Fisher 2013).

Terrorists may also use Twitter for practical communication. When US airstrikes against Syria seemed imminent in August 2013, several jihadi and Hezbollah-related groups in Syria used Twitter's real-time function to exchange urgent communication in order to prepare for attacks that they thought would be aimed at them. Some experts also consider it realistic that Twitter could be used to carry out actual attacks. In fact, in 2008 alone, an intelligence report released by the US Army's 304th Military Intelligence Battalion included a chapter entitled "Potential for Terrorist Use of Twitter," which expressed the army's concern over the use of the blogging services (304th MI BN OSINT Team 2008). The report says that Twitter could become an effective coordination tool for terrorists trying to launch militant attacks. It also highlights three possible scenarios of terrorist usage of this online format. In the first scenario, terrorists would send and receive near-real-time updates on the logistics of troops' movements in order to conduct more successful ambushes. In the second scenario, an operative with an explosive device or suicide belt would use a mobile phone to send images of his or her location to a second operative, who would use the near-actual-time imagery to time the precise moment to detonate the explosive device. In the third scenario, a cyberterrorist operative finds a soldier's online social media account, and is able to hack into his account and communicate with other soldiers using the stolen identity (304th MI BN OSINT Team 2008, 9). These scenarios illustrate the varied potential uses of Twitter to improve the accuracy and deadliness of terrorist attacks.

As terrorists increasingly have embraced social media, Twitter has faced criticism for hosting terror feeds. For instance, in February 2013 the British House of Commons summoned executives from major new media companies such as Twitter and Facebook to question them about how they plan to address this problem. Like most social media companies, Twitter has legal codes that prohibit violent and hate contents. According to Twitter's terms of use, users are prohibited from publishing or posting "direct, specific threats of violence against others." On the one hand, Twitter maintains a rigorous free-speech approach and repeatedly has refused to delete anti-Semitic, anti-Islamic, or other offending contents. For quite a while, it continued to tolerate al-Manar, Hezbollah's media wing, as a user.[3] On the other hand, Twitter has shut down numerous accounts that violated its terms of use; however, this approach has not always proved effective. When the Syrian Al-Nusra Front's Twitter account was suspended, it opened up an alternative account on August 25, 2013—which gained more than 20,000 followers within one day! This example illustrates how fast terrorists can restore the status quo ante online.

Besides the risk of shutdown, popular Twitter accounts may become a double-edged sword for terrorists. In 2013, Hezbollah's official account, on which the organization posted an average of more than 200 tweets a day, was hacked and suddenly tweeted "news" in favor of the Syrian rebels, such as "Sheikh Naim Qassem made a visit to Al-Qusayr city to raise the morale of the resistance's brave soldiers" (quoted in Stalinsky 2013a). Another example of a rather creative countermeasure is the "Twitter War" waged between the Taliban and the media department of the international coalition troops in Afghanistan (ISAF). After following the accounts of two Taliban spokesmen for intelligence reasons, in September 2011 coalition soldiers decided to directly post counternarratives to the propaganda. For instance, when @ABalkhi (Abdulqahar Balkhi, a Taliban spokesperson) tweeted that "@isafmedia continue genocide of Afghans: ISAF terrorists beat defenseless

---

[3] This case illustrates a relevant legal characteristic of new media. Although it is strictly forbidden for US companies to do business with or provide services for designated terrorist organizations, social media services like Twitter and Facebook are entirely passive; they do not receive any pay and do not provide customized services. Therefore, this law may not apply (Guttman 2012).

man to death," ISAF replied: "Sorry @ABalkhi: looting and beating inno-cents are NOT part of ISAF practices during routine searches" (Londono 2011). Since neither side can provide real evidence for their respective claims via Twitter, it is questionable whether this dialogue has any favorable impact on the audience. But according to the ISAF spokespersons, the Taliban have since refrained from tweeting overly exaggerated propaganda claims.

YouTube Terrorism

On the password-protected al-Qaeda online forum Al-Faloja, a posting states: "YouTube is among the most important media platforms in support-ing the mujahideen, as it is ranked third in the world with more than 70 mil-lion daily visitors."[4] YouTube was established in February 2005 as an online repository facilitating the sharing of video content. YouTube claims to be the "the world's most popular online video community." According to YouTube, on average, more than one billion users watch about six billions hours of videos every month, and 100 new hours are uploaded every minute. Over-all, YouTube passed 1 trillion watched videos in 2011; statistically speak-ing, that means 140 views for every human being on the planet. YouTube is localized in 56 countries and featured in 61 languages, and 70 percent of YouTube traffic comes from outside the United States (YouTube 2014).

The gigantic video sharing website has become a significant platform for jihadist groups and supporters, fostering a thriving subculture of jihadists who use YouTube to communicate, share propaganda, and recruit new indi-viduals to the jihadist cause. As Rita Katz and Josh Devon (2014) revealed:

> Easily accessible from almost anywhere in the world, YouTube's mas-sive global audience ensures that jihadists can simultaneously target both potential recruits as well as those whom the movement intends to ter-rorize. Furthermore, rather than having to wait for an extended period of time to download videos from jihadist forums, users on YouTube can watch virtually instantaneously. . . . As important as the videos hosted on YouTube, though, is the website's facilitation of social networking among jihadists. Comments left on videos and user channels, as well as the capability to send private messages to other users, helps jihadists

---

[4] Posted on the Al-Faloja forum, http://www.al-faloja.info/vb/showthread.php?t=62982.

identify each other rapidly, resulting in a vibrant jihadist subculture on YouTube. This community is comprised of many of the same individuals active on jihadist forums, who create their own video channels of jihadist propaganda designed to cultivate an atmosphere that radicalizes others.

Terrorist groups realized the potential of this easily accessed platform for disseminating propaganda and radicalization videos. Terrorists themselves praised the usefulness of this new online apparatus: "A lot of the funding that the brothers are getting is coming because of the videos. Imagine how many have gone after seeing the videos. Imagine how many have become *shahid* [martyrs]," convicted terrorist Younis Tsouli (who went by the user name "Ithabi007" ["Terrorist007"]) testified in his 2007 trial (quoted in Weimann 2010a, 51).

In 2008, jihadists suggested a "YouTube Invasion" to support jihadist media and the administrators of al-Fajr-affiliated forums. This suggestion was posted on the Al-Faloja forum on November 25, 2008. The posting provides a synopsis of the YouTube site and its founding, and notes its use by US president-elect Barack Obama during his presidential campaign, and by others. YouTube is argued to be an alternative to television and a medium that allows for jihadists to reach massive, global audiences. The message even instructs how jihadists should cut mujahideen videos into 10-minute chunks, as per YouTube's requirements, and upload them sequentially to the site. "I ask you, by Allah, as soon as you read this subject, to start recording on YouTube, and to start cutting and uploading and posting clips on the jihadist, Islamic, and general forums," said the posting on the jihadist website ("Jihadist Forum" 2008). "Shame the Crusaders by publishing videos showing their losses, which they hid for a long time." The jihadists themselves noted the success of the "YouTube Invasion" in the following statement posted on the Al-Faloja forum in December 2008:

> After the great success accomplished by the YouTube Invasion and the media uproar it caused that terrorists are getting trained to use YouTube, we have to clarify some matters about the YouTube Invasion. It is a continuous and successive invasion of YouTube. It does not have a time frame or that it will be over after a while. No, it is ongoing and flowing. Brothers, you have to study YouTube in a detailed manner, because it will be one of the pillars of jihadist media. It will be a permanent tool, Allah willing, to reveal the Crusaders and their helpers. Strive to discover new tools and means . . . .

Among the more widely known jihadist distributors on YouTube is Colleen LaRose, known as "Jihad Jane," who in March 2009 was charged with conspiring to provide material support to terrorists. LaRose maintained several YouTube channels replete with jihadist content. According to her indictment, LaRose posted on YouTube that she was "desperate to do something somehow to help" the plight of Muslims. Even more influential was Anwar al-Awlaki, the American-born radical cleric killed in Yemen in 2011, whose 700 YouTube videos have been watched more than 3.5 million times. Because of the popularity of his blog, Facebook page, online magazine *Inspire*, and hundreds of YouTube videos, the Saudi news station Al Arabiya has described al-Awlaki as the "bin Laden of the Internet" (quoted in Morris 2010). A 2009 British government analysis of YouTube found 1,910 videos of al-Awlaki, one of which had been viewed 164,420 times (Gardham 2010). In one example of al-Awlaki's effect through YouTube, his online videos inspired Roshonara Choudhry, a Muslim student jailed for attempting to murder British member of Parliament Stephen Timms in May 2010. Choudhry claimed that she was radicalized after being directed inadvertently by YouTube to a stream of al-Awlaki's videos (Dodd 2010). Despite his death and the demands to remove his videos, as of January 1, 2014, a simple keyword search for "Al Awlaki lectures" brought up more than 300 YouTube video clips. In one Arabic-language YouTube video, for example, al-Awlaki told viewers: "Don't consult with anybody in killing the Americans, fighting the devil doesn't require consultation or prayers seeking divine guidance. They are the party of the devils."

In recent years, terrorist clips—many of which correlate to major terrorist events around the world—have proliferated on YouTube. The following are some notable examples:

- Following the March 29, 2010, Moscow subway bombing carried out by two Chechen women, Chechen rebel leader Doku Umarov claimed responsibility for the attack in a video posted directly on YouTube. The video was announced on the website of the Kavkaz Center, a privately run portal associated with Chechen jihadist movement.
- On April 30, 2010, the Tehreek-e-Taliban Pakistan group created its official YouTube page. One day later, on May 1, the group posted its first video on that page—a video that claimed responsibility for that day's attempted car bombing attack in New York City's Times Square.
- During 2012, numerous YouTube pages posted installments from a series of detailed video lessons produced by the Izz al-Din al-Qassam

Brigades, the military wing of Hamas. The series was titled *Waa'iddu* ("And Prepare"), a Hamas slogan taken from a Qur'anic verse (8:60) that instructs Muslims to prepare for battle with the enemy. Most of these videos were about half an hour long, and discussed techniques for making bombs and using various weapons.

Some terrorist groups have even launched their own YouTube or self-controlled versions of it. In 2008, Hamas launched AqsaTube, its own video-sharing website, and described it as "the first Palestinian website specializing in Islamic and jihad audio-visual productions." In addition to choosing a similar name, the logo was designed to look just like YouTube's logo and page design. AqsaTube featured content ranging from relatively conservative Syrian social drama to videos glorifying al-Qaeda, the latter including commemoration of *shaheeds* (martyrs), songs, and the glorification of operatives from the military Izz al-Din al-Qassam Brigades. Once certain Internet providers refused to host the website, Hamas launched newer versions, named PaluTube and TubeZik. The Sri Lankan Tamil nationalist group Tamil Tigers has also launched TamilTube. In addition to providing alternatives to YouTube, terrorists have taught methods to bypass YouTube's context restrictions and regulations. In a series of postings starting on April 19, 2011, members of the password-protected English-language Ansar al-Mujahideen forum shared strategies for evading YouTube censorship of materials promoting jihad. These postings suggested various techniques for rendering jihadist content less conspicuous to YouTube administrators and individuals who search for jihadist content for flagging, and also gave strategies to make it easier to create new accounts and to minimize the loss of contacts and information after accounts are removed (SITE Monitoring Service 2011b).

A study conducted by Maura Conway and Lisa McInerney (2008) analyzed the online supporters of jihad-promoting video content on YouTube, focusing on those posting and commenting upon martyr-promoting material from Iraq. The findings suggest that a majority are under 35 years of age and reside outside the Middle East and North Africa region, with the largest percentage of supporters located in the United States. As the researchers concluded, "What is clearly evident, however, is that jihadist content is spreading far beyond traditional jihadist websites or even dedicated forums to embrace, in particular, video sharing and social networking—both hallmarks of Web 2.0—and thus extending their reach far beyond what may be conceived as their core support base in the Middle East and North Africa region to

Diaspora populations, converts, and political sympathizers" (Conway and McInerney 2008, 11).

Similar to the other leading social media, YouTube forbids any content that would be regarded as an incitement to violence. YouTube has also responded to numerous government requests to remove videos of radical groups. Despite these efforts, many videos remain available, and even more terrorist propaganda is constantly being posted. As terrorists pepper YouTube with a constant stream of new videos, it is difficult for YouTube to "flag" and delete objectionable content.[5] Often, it will take months for an offensive video to be removed from the site. It is also common for any visitors that flag offensive content to be bombarded with abusive and violent responses from jihadist supporters. YouTube's terms of service states that videos "inciting others to commit violent acts" can be flagged, and viewers should notify YouTube of such content. However, an experiment conducted in 2013 revealed that the effectiveness of the flagging system is questionable: Out of 125 videos flagged, 57 (45.4%) were still online more than four months later (Stalinsky and Zweig 2013). Finally, if speech "which attacks or demeans a group" is barred according to YouTube guidelines, why does a YouTube search of "kill the infidels" yield 1,052 hits?

In conclusion, the Internet has proved to be a useful instrument for modern terrorists who use it for a wide range of purposes, from data mining and online fund-raising to incitement, recruitment, and propaganda. However, in this new virtual battlefield, terrorists have needed to renovate, change, and update their online presence. Because counterterrorism agencies have shut down major terrorist websites and forums (and many such websites are now disrupted daily, with users experiencing increasing difficulties in posting material on these websites), the new terrorists turn to newer online platforms. There is a clear trend of terrorist "emigration" to online social media including YouTube, Twitter, and Facebook. Moreover, this trend is expanding to the newest online platforms, such as Instagram, Flickr, and others. For "newcomers," converts, followers, or sympathizers of jihadi or other terrorist movements, the threshold to access this illegal (or at least illegitimate) content on mainstream new media pages is much lower than to sign up for the closed hardcore forums. Rephrasing von Clausewitz's classic expression, the new media should be regarded as "an increasing continuation of war

---

[5] According to YouTube, users upload more than 72 hours of video to YouTube every minute. With so much content on the site, it would be impossible to review it all. As a result, YouTube relies on its users to flag content that they find inappropriate and let YouTube delete it.

by other means." The new arena of cyberspace, with its numerous online platforms, presents new challenges and requires dramatic shifts in strategic thinking regarding national security and counterterrorism. The strategic thinking should, for example, plan for the future: Since terrorists are so skilled and successful in adapting new communication technologies, counterterrorist approaches should consider future developments, emerging new platforms, and ways to anticipate and preempt terrorist abuse of these tools.

Part III

Future Threats and Challenges

In this dynamic, ever-changing, and constantly modifying cyber environment, predicting future online terrorist threats and challenges is very risky and almost impossible. Yet, at least in the short run, several emerging trends can indicate what to expect in the near future. Online terrorist "chatter" is another source for predicting future trends. For more than 15 years, our research team has been monitoring terrorist websites, forums, chatrooms, social media, and other online platforms. This "chatter" can provide information on terrorists' thought patterns; their optimal strategies, tactics, and modes of operation; and their true intentions. It can also be a source of learning about desired targets, recommended instructional material such as manuals, and new weapons under consideration. This online information has proved to be useful in numerous cases where attacks either failed or were prevented, as well as in identifying changes in terrorist methods of operation (e.g., the use of "econo-jihad" [Weimann 2009c]), cyberattacks, recruitment of women and children). Combining these sources have helped to develop the predictions made in this part that relate to both future terrorist plans and potential online counterterrorism measures.

Chapter 8

Cyberterrorism

Tomorrow's terrorist may be able to do more damage with a keyboard than with a bomb.

—National Research Council, *Computers at Risk* (1991)

Some years ago, in a television interview, I compared the threat of cyberterrorism to a dark cloud on the horizon: it is there, it is dark, it is coming, and yet we do not know when it will arrive. Today, I know that the dark cloud is already here. As this chapter will reveal, there is no doubt that terrorists today are considering the use of cyberattacks—by attempting to recruit the personnel and develop the know how to do so—and some have already launched such attacks. Yet even though the dark cloud of cyberterrorism is already here, its full dreadful capacities are not. The September 11 attacks present a tough challenge for any terrorists who would like to surpass the magnitude and impact of that day's events. Cyberterrorism, from the terrorist perspective, is one of the most promising weapons to respond to this challenge.

What Is Cyberterrorism (and What Is Not)?

Cyberterrorism is commonly defined as the use of computer network devices to sabotage critical national infrastructures such as energy, transportation, or government operations. The premise of cyberterrorism is that as modern infrastructure systems have become more dependent on computerized networks for their operation, new vulnerabilities have emerged—"a massive electronic Achilles' heel" (Lewis 2002, 1).

The roots of the notion of cyberterrorism can be traced back to the early 1990s, when the rapid growth in Internet use and the debate on the emerging "information society" sparked several studies on the potential risks faced by the highly networked, high-tech-dependent societies. As early as 1990, the prototypical term "electronic Pearl Harbor" was coined, linking the threat of a computer attack to an American historical trauma. The term itself was first used in the 1980s by Barry Collin, who discussed this dynamic of terrorism as transcendence from the physical to the virtual realm and "the intersection, the convergence of these two worlds" (Collin 1997, 15). Although cyberterrorism combines the terms "cyber" and "terror," the full term "cyberterrorism" is an even more opaque and broadly defined term than "terrorism," adding another layer to an already contentious concept.

Cyber events in general are often misunderstood by the public and erroneously reported by the media. Several difficulties have prevented the creation of a clear and consistent definition of the term "cyberterrorism." First, much of the discussion has been conducted in the popular media, where journalists typically prefer drama and sensation to good operational definitions of new terms. Second, it has been especially common when dealing with computers to coin new words simply by placing variants of terms such as "cyber," "computer," or "information" before another word. Thus, a slew of terms—"cybercrime," "infowar," "netwar," "cyberterrorism," "cyberharassment," "virtual warfare," "digital terrorism," "cybertactics," "computer warfare," "cyberattack," and "cyber-break-ins"—is used to describe what some military and political strategists call "new terrorism" of our times.

Some efforts have been made to introduce greater semantic precision. Dorothy Denning, a professor of computer science, put forward an unambiguous definition in numerous articles and in her testimony on the subject before the House Armed Services Committee in 2000:

> Cyberterrorism is the convergence of cyberspace and terrorism. It refers to unlawful attacks and threats of attacks against computers, networks

and the information stored therein when done to intimidate or coerce a government or its people in furtherance of political or social objectives. Further, to qualify as cyberterrorism, an attack should result in violence against persons or property, or at least cause enough harm to generate fear. Attacks that lead to death or bodily injury, explosions, or severe economic loss would be examples. Serious attacks against critical infrastructures could be acts of cyberterrorism, depending on their impact. Attacks that disrupt nonessential services or that are mainly a costly nuisance would not (Denning 2000, 1).

The National Conference of State Legislatures, an organization of legislators created to help policymakers with issues such as economy and homeland security, gives the following definition for cyberterrorism:

> The use of information technology by terrorist groups and individuals to further their agenda. This can include use of information technology to organize and execute attacks against networks, computer systems and telecommunications infrastructures, or for exchanging information or making threats electronically. (Quoted in Gorge 2007)

It is important to distinguish the terms cyberterrorism, hacking, and "hacktivism." Hacking is understood here to mean activities conducted online and covertly that seek to reveal, manipulate, or otherwise exploit vulnerabilities in computer operating systems and other software. Most hackers tend *not* to have political agendas, and concentrate on writing programs that expose security flaws in computer software. Their efforts in this direction have sometimes embarrassed corporations but have also been responsible for alerting the public and security professionals to major software security flaws. Moreover, although hackers have been known to damage systems, disrupt e-commerce, and force websites offline, the vast majority of hackers do not have the necessary skills and knowledge to inflict serious harm, and the ones who have those skills generally do not seek to do so (Denning 2001).

Hacktivism is a related term coined by scholars to describe hacking with a political activism component. Although hacktivism is politically motivated, it does not constitute cyberterrorism. Hacktivists want to protest and disrupt; they do not want to kill, maim, or terrorize. However, hacktivism does highlight the threat of cyberterrorism: individuals with no moral restraint may use methods similar to those developed by hackers to wreak havoc. The line between cyberterrorism and hacking or hacktivism may blur, especially if terrorist groups recruit or hire computer-savvy hacktivists or if hacktivists decide to escalate their actions by attacking the systems that operate critical elements of the national infrastructure, such as electric power networks and emergency services (Weimann 2005a, 136–37).

Michael Vatis (2001) has classified potential cyberattackers in four categories:

- *Terrorists:* Although only few terrorist groups have used cyberterrorism, many have shown interest in using it and are attempting to master it.
- *Nation-states:* Several nation-states, including supporters of terrorism (Syria, North Korea, Iran, Sudan, and Libya), have developed cyberwarfare capabilities and even employ them.
- *Terrorist sympathizers:* Various hacker groups have the ability to launch cyberattacks to show their support for a terrorist group or its cause. Vatis finds that these groups are the likeliest to engage in cyberterrorism.
- *Thrill-seekers* (or "cyber-joyriders"): According to Vatis (2001, 14), these hackers and "script kiddies" (a derogatory term used to describe individuals who attempt to break the security on computer systems without understanding the exploits they are using) simply want to gain notoriety through high-profile attacks. However, such individuals can and have had significant disruptive impact through their destructive attacks.

Vatis's list provides a basis for classifying potential cyberattackers, though it does not include other categories of cyberattackers, including criminals who engage in extortion, identity theft, credit card and bank fraud, and corporate espionage; and "insiders" who engage in sabotage, fraud, and so on in systems to which they already have access.

The Appeal of Cyberterrorism

Cyberterrorism is an attractive option for modern terrorists for several reasons:

- *Minimal resources required:* Cyberterrorism is cheaper than traditional terrorist methods. All that the terrorist needs is a personal computer and an online connection. Terrorists do not need to buy weapons such as guns and explosives; instead, they can launch digital attacks through a telephone line, a cable, or a wireless connection.

The minimal resources that are needed for such an attack—one person in front of a computer connected to the Internet—helps groups that have limited funds.

- *Anonymity:* Cyberterrorism is more anonymous than traditional terrorist methods. Like many Internet surfers, terrorists use online nicknames—"screen names"—or log on to a website as an unidentified "guest user," which makes it harder for security agencies and police forces to track down their real identity. Cyberspace also has no physical barriers to overcome, such as checkpoints to navigate, borders to cross, or customs agents to outsmart.
- *Remote attacks:* Cyberterrorism can be conducted remotely, a feature that is especially appealing to terrorists. In fact, a terrorist at a computer on one side of the world can launch an attack, route it through dozens of different countries, cover his tracks so that it is nearly untraceable, and cause great damage to a society or nation on the other side of the globe.
- *Vulnerabilities:* The variety and number of targets are enormous. Cyberterrorists can target the computers and computer networks of governments, individuals, public utilities, private airlines, and so forth. The sheer number and complexity of potential targets guarantee that terrorists can find weaknesses and vulnerabilities to exploit. Several studies have shown that critical infrastructures such as electric power grids and emergency services are vulnerable to a cyberterrorist attack because both the infrastructures and the computer systems that run them are highly complex, making it effectively impossible to eliminate all weaknesses.
- *Scope of damage:* The potential scope of damage is another attractive feature of cyberterrorism. Consider the following comparative scenario: A suicide bomber can enter a bus, and if successful can manage to kill all the passengers on the bus and possibly harm bystanders and others in the immediate area. With cyberattacks, a terrorist can take control of traffic lights in a certain area, the air traffic control systems of a busy airport, or the computers controlling the underground transport system in a major city—and could cause hundreds or even thousands of fatalities over a much wider area.[1]

---

[1] For example, in December 2013, Jeff Kohler, vice president of international business development for Boeing's defense arm, admitted to being "very concerned" about threats to aviation software and said that aircraft were now in need of cyber protection. Planes are at risk every time they enter an airport because of the number of electronic systems they begin sharing information with, and the situation will cause "a lot of issues" in the coming years, he added (quoted in Collins 2013).

- *Greater "fear factor":* Cyberterrorism fits with terrorists' goals of infusing fear into the lives of their enemies. Cyberterrorism can come without any warning, and there is not much that ordinary civilians can do to protect themselves against such attacks. This uncertainty and lack of control over one's own world make the prospect of this form of terrorism such a dreadful option.

These factors are all reasons why cyberterrorism is a much more appealing form of attack. Terrorists are indeed getting interested in cyberwarfare. In his March 2012 testimony before a US House of Representatives appropriations subcommittee, Federal Bureau of Investigation (FBI) director Robert Mueller said that terrorists may seek to train their own recruits or hire outsiders with an eye toward pursuing cyberattacks on the United States. "Terrorists have not used the Internet to launch a full-scale cyber attack, but we cannot underestimate their intent," Mueller said. He also stated that terrorists have shown interest in developing hacking skills, and that the evolving nature of the problem makes the FBI's counterterrorism mission more difficult (quoted in Associated Press 2012).

The Forms of Cyberterrorism

Even now, we are under cyberattack. Using statistics from the online community hackerwatch.org, in one week of March 2014, more than three million serious computer attacks were reported, with 443,552 such attacks taking place in one 24-hour period—a rate of more than 300 attacks per minute.[2] Though most of these attacks are not committed by terrorists, there is a growing share in the part that terrorism plays in cyberattack patterns.

Cyberattacks have various forms and categorizations. One such classification relies on the objectives of the attackers, which include the following four areas:

1. Loss of integrity, such that information could be modified improperly;
2. Loss of availability, where mission-critical information systems are rendered unavailable to authorized users;
3. Loss of confidentiality, where critical information is disclosed to unauthorized users; and

---

[2] HackerWatch reports information that helps identify, combat, and prevent the spread of Internet threats and unwanted network traffic (http://hackerwatch.org/; data retrieved March 20, 2014).

4. Physical destruction, where information systems create actual physical harm through commands that cause deliberate malfunctions. (Rollins and Wilson 2007, 3)

A fifth objective often mentioned is publicity, where even a marginally successful cyberattack directed at a major facility or service is sufficient to garner considerable publicity and consequently increases public anxiety and distrust.

The cyberattack on Estonia may illustrate the potential of a well-orchestrated cyberattack directed at a specific nation (Landler and Markoff 2007). In May 2007, several key Estonian government and business computer systems were subjected to a mass cyberattack following the controversial removal of a Soviet-era World War II war memorial from downtown Tallinn. The attack was a distributed denial-of-service (DDoS) attack in which selected sites were bombarded with traffic in order to force them offline. The cyberattack affected nearly all Estonian government ministry networks as well as two major Estonian bank networks, all of which were knocked offline. Despite speculation that the attack had been coordinated by the Russian government, Estonia's defense minister admitted that he had no evidence linking the cyberattacks to the Russian authorities. NATO and the United States sent computer security experts to Estonia to help the nation recover from these cyberattacks, and to analyze the methods used and determine the source of the attacks. Some security experts suspect the involvement of cybercriminals, possibly using a large network of infected personal computers (called a "botnet"), to help disrupt the Estonian government's computer systems.

An attack against computers may disrupt equipment and hardware reliability, change processing logic, or steal or corrupt data (Wilson 2007; Wilson 2008). Various methods can be used for such attacks:

- Conventional kinetic weapons (e.g., firearms, explosives) can be directed against computer equipment, a computer facility, or transmission lines in a physical attack that disrupts the reliability of the equipment.
- Electromagnetic energy, most commonly in the form of an electromagnetic pulse, can be used to create an electronic attack directed against computer equipment or data transmissions. By overheating circuitry or jamming communications, electronic attacks disrupt equipment reliability and data integrity.

- A computer network attack (CNA), directed against computer processing code, instruction logic, or data, can generate a stream of malicious network packets intended to disrupt data or logic by exploiting vulnerability in computer software, or weaknesses in an organization's computer security practices.

Cyberterrorism is often equated with the last method, the use of malicious code. The 2007 Estonian case is one such example. However, a cyberterrorism event may also depend on the use of other measures. Thus, it is possible that if certain computer facilities were deliberately attacked for political purposes, all three methods described above (physical attack, electronic attack, and cyberattack) might contribute to or be labeled as "cyberterrorism." Where do vulnerabilities lie, and what technological tools will terrorists use? The following sections discuss some of the types of "cyber weapons" that terrorists have at their disposal.

Botnets

Botnets, or "bot networks," are made up of vast numbers of compromised computers that have been infected with malicious code and can be remotely controlled through commands sent via the Internet. Hundreds or thousands of these infected computers can operate in concert to disrupt or block Internet traffic for targeted victims. Once the botnet is in place, it can be used in DDoS attacks, proxy and spam services, malware distribution, and other organized criminal or terrorist activity. Botnets can also be used for covert intelligence collection or to attack Internet-based critical infrastructure. Additionally, botnets can be used as weapons in propaganda or psychological campaigns against their targets to instigate fear, intimidation, or public embarrassment. Botnets are becoming a major threat for future cyberterrorism, partly because they can be designed to disrupt targeted computer systems in different and effective ways, and because even terrorists that do not have strong enough technical skills to develop their own botnets can apply these disruptive measures in cyberspace simply by renting botnet services from a cybercriminal.[3] According to a June 2013 FBI report, the use

---

[3] For example, Jeanson Ancheta, a 21-year-old hacker and member of a group called the "Botmaster Underground," reportedly made more than $100,000 from different Internet advertising companies who paid him to download specially designed malicious adware code onto more than 400,000 vulnerable PCs he had secretly infected and taken over. He also made more money by

of botnets is on the rise, and it estimated that "botnet attacks have resulted in the overall loss of millions of dollars from financial institutions and other major U.S. businesses. They've also affected universities, hospitals, defense contractors, law enforcement, and all levels of government" (Federal Bureau of Investigation 2013).

Botnet codes for infecting computers were originally distributed as infected email attachments, but additional methods can be used to acquire more computers for the system. A website may be unknowingly infected with malicious code in the form of an ordinary-looking advertisement banner, or the site may include a link to an infected website. Clicking on the banner or following the link may install botnet code. Botnet codes can also be silently uploaded to a user's computer simply by exploiting an unpatched security vulnerability in the user's Internet browser—even if the user takes no action while viewing the website. Some bot software can even disable the user's antivirus security before infecting the computer. Once infected, the malicious software establishes a secret communications link to a remote "botmaster" in preparation to receive new commands to attack a specific target (Wilson 2008).

Attacks on SCADA Systems

SCADA (supervisory control and data acquisition) is a type of computer-controlled system that monitors and controls systems such as industrial, infrastructure, and facility-based processes. SCADA is one part of the broader category of industrial control systems, which include programmable logic controllers, remote terminal units, and other monitoring and automation devices used in all types of industrial, infrastructure, and facility processes and systems. Industrial processes that use SCADA systems include manufacturing, production, power generation, fabrication, and refining. Infrastructure processes may be public or private, and include water treatment and distribution, wastewater collection and treatment, oil and gas pipelines, electrical power transmission and distribution, wind farms, civil defense siren systems, and large communication systems. Facility processes involve monitoring and controlling heating, ventilation, and air conditioning systems and energy consumption in buildings, airports, ships, and space stations.

---

renting his 400,000-unit "botnet herd" to other companies that used it to send out spam, viruses, and other malicious code. In 2006, Ancheta was sentenced to five years in prison (Wilson 2008).

SCADA systems have existed since the 1960s. In the early days, they were stand-alone, and few were networked. Today, virtually all are accessed via the Internet. This technological development has helped to cut costs, but from an information security perspective it introduces vulnerabilities. SCADA systems that tie together decentralized facilities such as power, oil, gas pipelines, and water distribution and wastewater collection systems were designed to be open, robust, and easily operated and repaired, but not necessarily secure. Alarmingly, in 1997 the President's Commission on Critical Infrastructure Protection said of SCADA systems:

> From the cyber perspective, SCADA systems offer some of the most attractive targets to disgruntled insiders and saboteurs intent on triggering a catastrophic event. With the exponential growth of information system networks that interconnect the businesses, administrative and operational systems, significant disruption would result if an intruder were able to access a SCADA system and modify the data used for operational decisions, or modify programs that control critical industry equipment or the data reported to control centers. (President's Commission on Critical Infrastructure Protection 1997, A-27)

Advancements in the availability and sophistication of malicious software tools have increased the cyber threat to these systems, as new technologies raise new security issues that cannot always be addressed prior to adoption. The increasing automation of critical infrastructures provides more cyber access points for terrorists to exploit. As was described in the October 2005 joint hearing before the Subcommittee on Economic Security, Infrastructure Protection and Cybersecurity:

> Securing SCADA systems is similar to securing all of our cyber infrastructure; however, the consequences are potentially very different. Minimally, adversaries could target SCADA systems through cyber networks, utilizing common cyber attack methods to render the SCADA systems unusable. This could slow down, stop, or endanger the functions of the facility. This would result in not only serious problems at that facility but potential cascading effects on other facilities or processes that are dependent on the attacked facility. Even worse, terrorists could utilize SCADA systems for their own sinister motives—causing a pipeline to burst, opening flood gates on dams, or shutting down our electric supply, all without ever gaining access to the facility. ("SCADA Systems and the Terrorist Threat" 2005, 2)

All of our infrastructure systems rely on computers. Most of those computers may be especially vulnerable, and their importance for controlling the critical infrastructure may make them an attractive target for cyberterrorists. SCADA systems, once connected to isolated networks using only proprietary computer software, now operate using more vulnerable commercial off-the-shelf software and are increasingly being linked directly to corporate office networks via the Internet. Many experts believe that most SCADA systems are inadequately protected against a cyberattack and remain persistently vulnerable because many organizations that operate them have not paid proper attention to their unique computer security needs (Wilson 2005, 10).

Denial-of-Service Attacks

Cyberterrorists may also use denial-of-service attack methods to overburden the computers of a government and its agencies. Denial-of-service (DoS) attacks are designed to make a computer or network of computers unavailable to its users. One common method of attack involves saturating the target machine with external communications requests, to the point that it cannot respond to legitimate traffic or responds so slowly as to be rendered essentially unavailable. On a networked computer, such attacks usually overload the server and affect all of its users.

If an attacker uses a single host to launch the attack, this approach is classified as a DoS attack. However, if an attacker uses the capabilities of many systems (such as botnets) to launch simultaneous attacks against another host, this is classified as a DDoS attack. For this purpose, the attack can use viruses or other malware to infect several unprotected computers and then take control of them. Once control is obtained, the terrorists can manipulate these infected computers to initiate the attack, such as by using botnets to send information or demand information in such large numbers that the victim's server effectively collapses under the strain of processing the information. A stronger version of this attack, known as permanent denial-of-service (PDoS), can even damage a system so badly that the system's hardware must be reinstalled or even replaced. Unlike a DDoS attack, which attempts to overload a system, a PDoS attack exploits security flaws that enable the attacker to gain control over the victim's hardware, such as routers, printers, or other networking hardware. The attacker uses these vulnerabilities to modify, corrupt, or install defective firmware to the victim's system—a process which, when done legitimately (such as to upgrade the device), is known as flashing. The corrupted firmware "bricks" the device, rendering it unusable for its original purpose and requiring the victim to repair or replace it, often at great expense.

DDoS attacks became more commonly known in early 2000, when attackers managed successful strikes against popular websites such as CNN, Yahoo!, and Amazon. Although many years have passed since they first appeared in the mainstream, DDoS attacks are still difficult to block. Indeed, some DDoS attacks can be impossible to stop if they have sufficient resources behind them. It is estimated that at least 50 percent of Fortune 500 companies have been compromised by such attacks, and the potential financial damage to these organizations is almost impossible to quantify, but it is probably in the trillions of US dollars (Armerding 2012). In September 2012, DDoS attacks shut down the websites of Bank of America and JPMorgan Chase and crippled those of Wells Fargo, U.S. Bank, and PNC Bank. The Hamas-affiliated Islamist group Izz ad-Din al-Qassam Cyber Fighters publicly claimed responsibility for the attacks in what it called "Operation Ababil." The group has launched attacks in the past, albeit ones that were far less coordinated than their 2012 success. As a report on these attacks concluded: "No matter who is behind the attacks, if a terror group can so easily crash a major banking website, what's next? Government systems like air traffic control? Or, critical infrastructure targets such as power grids? The prospects are mind-numbing, and frankly, scary" (Rothman 2012).

In April 2013, a jihadist who went by the name "Abu Obeida al-Masri" posted a videotaped tutorial for a program he developed to facilitate DDoS attacks against "Zionist-Crusader" websites, and invited fellow al-Qaeda supporters to join the "Electronic Islamic Army." Al-Masri asked members to join the Electronic Islamic Army's Facebook page and download the DDoS program, explaining that while such attacks are old-fashioned and simple, they are effective and difficult to stop. On the forum, Al-Masri gave detailed instructions with pictures how to use the program, and on April 21 he uploaded a video tutorial to YouTube in which he used Facebook as an example of a targeted site. He told supporters: "I pray to Allah that this work be for Allah's countenance and to benefit us all, and to make you and I a reason for the removal of the nation of disbelief, and to make us and you a thorn in the throat of the disbelievers and their helpers from among the tyrannical apostates. I remind you: Determination, determination, words for actions, and you are only responsible for yourself. Choose for yourself a field so that Allah will make you one of the knights. Don't neglect action, but neglect discouragement and sitting down. Wait for everyone to react to the action. We don't want slogans; instead, our slogan is action" (SITE Monitoring Service 2014b).

Cyber 9/11? The Likelihood of Cyberterrorism

The cyber terrorism threat is real, and it is rapidly expanding.

—FBI Director Robert S. Mueller III, Cyber
Security Conference, March 4, 2010

On January 29, 2014, Director of National Intelligence James R. Clapper Jr. presented the 2014 annual US intelligence community worldwide threat assessment in congressional testimony. In the published report, Clapper provided a thorough review of the status of possible threats from a wide variety of nations and terror groups. The report highlighted that critical cyber threats are converging:

In the past several years, many aspects of life have migrated to the Internet and digital networks. These include essential government functions, industry and commerce, health care, social communication, and personal information. . . . We assess that computer network *exploitation* and *disruption* activities such as denial-of-service attacks will continue. Further, we assess that the likelihood of a *destructive* attack that deletes information or renders systems inoperable will increase as malware and attack tradecraft proliferate (Clapper 2014 [emphasis in original]).

In the past, it was assumed that although terrorists were adept at spreading propaganda and attack instructions on the Internet, their capacity for offensive computer network operations was limited. Thus, in 2009 the FBI reported that cyberattacks attributed to terrorists were largely limited to unsophisticated efforts such as email bombing of ideological foes, DoS attacks, or defacing of websites.[4] However, the FBI report also noted that terrorists' increasing technical competency could result in an emerging

---

[4] Statement of Steven Chabinsky, Deputy Assistant Director, FBI Cyber Division, before the Senate Judiciary Committee Subcommittee on Homeland Security and Terrorism, at the *Cybersecurity: Preventing Terrorist Attacks and Protecting Privacy Rights in Cyberspace* hearing, November 17, 2009.

capability for network-based attacks. The FBI predicted that terrorists will either develop or hire hackers to complement future large conventional attacks with cyberattacks: "As shocking as 9/11 was to the nation, it was only a small breach compared to the systemic threats we face today," said former National Security Agency director Michael McConnell in a 2009 interview. "When the terrorists get smarter, they won't even need to come to our shores to create the kind of havoc and turmoil they did by flying planes into the Twin Towers. They will be able to do it from their laptops from overseas" (Gardels 2009). Clapper's 2014 remarks highlight the shift in thinking since the FBI's report five years earlier.

Continuing publicity about computer security vulnerabilities may encourage terrorists' interest in attempting cyberattacks. Take, for example, the case of Stuxnet. The threat of terrorist cyberattack became more plausible after the 2009 discovery of Stuxnet, a powerful computer worm used to attack Iran's nuclear program. The worm damaged Iran's nuclear centrifuges by causing them to spin too fast, which gave false information to the plant operators. The worm's creator has not been officially identified, though reports have alleged that the United States and Israel were behind the attack. Stuxnet could have a similar effect on other targets, and US officials expressed concern that the worm could be used by terrorists and their supporters. The US Department of Homeland Security (DHS) told Congress that it feared that the same attack could now be used against critical infrastructures in the United States and that the DHS "is concerned that attackers could use the increasingly public information about the code to develop variants targeted at broader installations of programmable equipment in control systems. Copies of the Stuxnet code, in various different iterations, have been publicly available for some time now" (Zetter 2011).

The Iranian nuclear program has not been the only victim of malicious computer code. In April 2012, cyberterrorists used a deadly computer virus to attack the information network of Aramco, the Saudi oil company. The virus, which annihilated all of the data on 35,000 desktop computers, also displayed the image of a burning American flag on the screens of the infected computers. A group called the Cutting Sword of Justice claimed credit for the attack (Dorgan 2013). In 2012 alone, NATO suffered around 2,500 cyberattacks on its networks, according to the alliance's secretary general (Farmer 2013). In March 2013, American Express customers trying to gain access to their online accounts were met with blank screens or an "ominous ancient type face" (Perlroth and Sanger 2013). The company confirmed that its website had been attacked. The assault was the latest in an intensifying campaign of unusually powerful attacks on American financial institutions that have taken dozens of them offline intermittently, costing millions of dollars. Similar attacks took JPMorgan Chase offline and incapacitated 32,000 computers at South Korea's banks and television networks (Perlroth and Sanger 2013).

Some terrorist hackers have moved beyond attacks on government and corporate targets and set their sights on online media. In May 2013, computer hackers hijacked the Twitter account of the Associated Press and sent a tweet stating that there had been two explosions at the White House and that President Barack Obama was injured. Within two minutes, the stock market dropped by 143 points. The Syrian Electronic Army later claimed credit for the attack. In August 2013, media companies including Twitter, the *New York Times*, and the *Huffington Post* lost control of some of their websites after hackers supporting the Syrian government breached the Australian Internet company that manages many major site addresses. Tweets from the Syrian Electronic Army claimed credit for the Twitter and *Huffington Post* attacks, and electronic records showed that NYTimes.com, the only site with an hours-long outage, redirected visitors to a server controlled by the Syrian group before it went down (Shih and Menn 2013).

Recent considerations of cyberterrorism have stressed the seriousness of the problem. As Meg King, the national security adviser to Jane Harman, director, president, and CEO of the Woodrow Wilson International Center for Scholars, argued:

Many information technology experts suggest that terror groups aren't now—and might never be—capable of carrying out an act of cyberterror. . . . But recent plots and propaganda suggest that the motive exists and the know-how is growing. Cyberterror is just around the corner: It could be a physical attack on the Internet's infrastructure, as attempted in London in 2007, that could halt important financial traffic. Or it might be an attack on a system controlling critical infrastructure—from oil refineries and nuclear plants to transportation networks. And we aren't prepared. (King 2014)

If terrorists want to surpass the magnitude and impact of 9/11, it seems that only a catastrophic cyberattack will be their option. Are they aware of it, and are they interested in launching this attack?

The Growing Interest of Terrorists in Cyberattack

"Hacking on the Internet is one of the key pathways to Jihad, and we advise the Muslims who possess the expertise in the field to target the websites and the information networks of big companies and government agencies of the countries that attack Muslims, and to focus on the websites and networks that are managed by the media centers that fight Islam, Jihad, and mujahideen."

—Al-Qaeda video, "You Are Held Responsible Only
for Thyself—Part 2," posted online on June 3, 2011

It is difficult to determine whether or which terrorist groups are capable of launching an effective cyberattack. However, there is growing evidence that modern terrorists are seriously considering adding cyberterrorism to their arsenal, as indicated by the widely cited statement by Frank Cilluffo of the Office of Homeland Security: "While bin Laden may have his finger on the trigger, his grandchildren may have their fingers on the computer mouse." Cyberterrorism expert Dan Verton, for example, argues that "al-Qaeda has shown itself to have an incessant appetite for modern technology" (Verton, 2003, 93) and provides numerous citations from Osama bin Laden and other al-Qaeda leaders that show their recognition of this new cyberweapon. In the wake of the September 11 attacks, bin Laden reportedly gave a statement to Hamid Mir of the Pakistan newspaper *Ausaf* indicating that "hundreds of Muslim scientists were with him who would use their knowledge . . . ranging from computers to electronics against the infidels" (quoted in Verton 2003, 108). Captured literature indicates that many al-Qaeda members are well educated and familiar with engineering and other technical areas (Spring 2004). In November 2001, when al-Qaeda fighters fled from a US attack on their base in Kabul, Afghanistan, they left behind documents and other information that exposed the degree to which some al-Qaeda operatives had been educated and trained in the use of computer systems (Davis 2002). One captured al-Qaeda computer contained engineering and structural architecture features of a dam, which had been downloaded from the Internet and would enable al-Qaeda engineers and planners to simulate catastrophic failures. US investigators also found evidence on other captured computers showing that al-Qaeda operators had spent time on sites that offer software and programming instructions for the digital switches that run power, water, transportation, and communications grids (Weimann 2008f).

Extremist groups that use and operate online platforms have also shown a significant increase in the level of their technical sophistication. In 2002, the Central Intelligence Agency (CIA) stated in a letter to the US Senate Select Committee on Intelligence that cyberwarfare attacks against the US critical infrastructure will become a viable option for terrorists as they become more familiar with the technology required for the attacks. Also according to the CIA, various groups (including al-Qaeda and Hezbollah) are becoming more adept at using the Internet and computer technologies, and these groups could possibly develop the skills necessary for a cyberattack (Verton 2003, 87). Later, FBI director Robert Mueller testified before the Senate Select Committee on Intelligence that terrorists show a growing understanding of the critical role of information technology in the US economy and have expanded their recruitment to include people studying math, computer science, and engineering.[5]

This technological familiarity encompasses more than simple computer know-how. A 2006 study of more than 200,000 multimedia documents on 86 sample websites concluded that extremists exhibited similar levels of web knowledge to US government agencies, and that the terrorist websites employed significantly more sophisticated multimedia technologies than US government websites (Qin et al. 2007). In 2010, "The Brigades of Tariq ibn Ziyad," a jihadist group with the stated goal of using cyber capabilities to penetrate US Army networks, launched a massive malware attack designed to impact businesses and government agencies. This particular attack was ideologically based, reinforced by the official video comment, "Listen to me about my reasons for the 9 September virus that affected NASA, Coca-Cola, Google, and most American [names]. What I wanted to say is that the United States doesn't have the right to invade our people and steal the oil under the name of nuclear weapons." Ominously, the creator of the video noted that the virus "wasn't as harmful as it could have been" (Greenberg 2010).

The monitoring and analysis of terrorist online chatter certainly reveal a growing interest in cyberattacks. In November 2011, a British government report on cybersecurity indicated that British intelligence had picked up "talk" from terrorists planning an Internet-based attack against the United Kingdom's national infrastructure. Indeed, the terrorist chatter reveals such interest: for example, a prominent jihadist not only suggested cyberattacks but also expressed interest in organizing a center for jihadists who have expertise in hacking, networking, and programming language. The jihadist,

---

[5] Testimony before the Senate Select Committee on Intelligence, February 16, 2005.

"Yaman Mukhdhab," posted his call for establishing an e-jihad center on the Shumukh al-Islam forum on June 11, 2011. Concerning cyberattacks, the posting highlights such attacks as a way to inflict massive damage to the economy of an enemy country, and noted that the United States is ill-prepared for an attack on its electrical grid, for example. Mukhdhab outlined the mission and requirements for the e-jihad center and stressed that only the "masters of disbelief"—France, the United Kingdom, and the United States—are to be targeted. He gave a priority list of targets in these countries, noting that SCADA systems that monitor industrial and infrastructure processes are at the top of the list, followed by systems that manage financial sites and companies, and sites in general that are connected with the "daily activities of the ordinary citizen." Mukhdhab provided forum members with a list of 19 categories for further study, including understanding SCADA systems, having fluency in machine and assembly languages, and having knowledge of websites frequented by hackers. He asked that they volunteer for only those categories in which they have expertise (Macdonald 2011).

The self-proclaimed Izz ad-Din al-Qassam Cyber Fighters, which successfully hit numerous major financial targets with DDoS attacks in September 2012, declared its future plans to continue its cyberterrorism campaign. In December 2012, the group announced that it will launch "phase 2" of its campaign to hack banking and financial websites, and named Bank of America, JPMorgan Chase, PNC, SunTrust, and US Bancorp as targets. The Cyber Fighters stated: "In [this] new phase, the wideness and the number of attacks will increase explicitly; and offenders and subsequently their governmental supporters will not be able to imagine and forecast the widespread and greatness of these attacks" (SITE Monitoring Service 2012a). Later, in a message posted on its Pastebin.com account on January 1, 2013, the Cyber Fighters reported that in the past few weeks of the second phase of its "Operation Ababil," it had attacked the websites of JPMorgan Chase & Co, Bank of America Corp, Citigroup Citibank, Wells Fargo & Company, US Bancorp, PNC Financial Services Group, BB&T Corporation, Suntrust Banks, and Regions Financial Corporation. The Cyber Fighters stated: "We, like most people in the United States, are banks' customers and we do not desire to disrupt the banks' financial transactions. But the American profiteer rulers' insistence and persistence in disregarding this reasonable demand of all Muslims of the world and not taking an action to remove this offensive film[6]

---

[6] The film in question is the controversial anti-Islamic short video *Innocence of Muslims*. In September 2012, the posting of this video online sparked demonstrations and violent protests in several Arab and Muslim nations and elsewhere around the world.

shows these tyrants insist that continue to insult Muslim saints. . . . So due to this irrational insistence on continuing the insults, it seems we should accustom ourselves to disruption in banking" (Kovacs 2013b).

In April 2013, a hacking group calling itself the "al-Qaeda Electronic Army" released a video threatening to attack America's "vital sectors" if the US government did not withdraw its soldiers from Muslim lands. The video, titled "Message from Ahmad bin Laden to the White House," was uploaded to YouTube (SITE Monitoring Service 2013a). A fellow hacking group, the "Tunisian Cyber Army," posted the video on its Facebook and Twitter pages and notified users that the threat is part of an upcoming operation dubbed "Black Summer" (Kovacs 2013a).

Terrorist Capabilities for Cyberattacks

The capability to launch cyberattacks against critical infrastructure using cyber resources is demonstrable and observable by looking at numerous past attempts and even successes. But do terrorists have the capabilities and the intent to apply the digital weaponry? As this chapter has revealed, terrorist organizations are realizing the value of the Internet both as a means of accomplishing their goals and as an objective in itself. In other words, the Internet can be seen as both a weapon and a target for cyberwarfare. In his analysis of existing jihadist cyberattack capabilities, Christopher Heffelfinger (2013, 1) argues that:

> The current pool of jihadist hackers is youthful, ambitious in its goals, and largely lagging in terms of its technical capabilities. This is best illustrated by the fact that these hackers have carried out few effective large-scale attacks to date. Jihadist hacktivists remain a loosely to [*sic*] disorganized set of individual hackers who form and disband hacking groups they create, and frequently enter into counterproductive rivalries with fellow hackers. Perhaps as a result, despite more than seven years of efforts to construct and recruit for jihadist hacking attacks via online forums, they have yet to form a jihadist hacking group that can demonstrably perform effective cyber attacks.

However, as Heffelfinger notes, jihadist-inspired hackers have a range of skillsets, leadership abilities, and hacking experience, and some of them have carried out small- to medium-scale cyberattacks against US government and private sector targets, with moderate impact in terms of data loss and exposure. Compared with hackers sponsored or controlled by state actors, jihadi hackers are clearly behind in terms of the impact of their attacks, their technical skillset, and their overall organizational and recruitment abilities. Their hacking activities frequently include website defacements, wherein the attackers leave antagonistic imagery and comments on the victimized websites. However, the activities of some jihadist hackers indicate that there is a gradual sophistication of attack modes and intended attack impacts, occurring alongside a growing contingent of young jihadist enthusiasts who see cyberattacks as an increasingly effective and relatively easy way to fight the West. As Heffelfinger (2013, 2) concludes,

> While jihadist-themed cyber attacks have been modest and often rudimentary over the past decade, the advancement and ambitions of certain jihadist hacking groups, individual hacktivists and proponents of cyber jihad over the past one to two years give some cause for concern in this area, particularly as those adversaries are growing more adept at identifying vulnerabilities in U.S. and other government targets, as well as those in the private sector. . . . The continuance of vulnerable attack targets and the likely increase in Islamist hacking activity in the near term combine to form a potentially challenging security environment for U.S. and other Western governments and private companies.

Many terrorist groups are reportedly building a massive and dynamic online library of training materials, many of which are supported by subject-matter experts who answer questions on message boards or in chatrooms. This online library covers many areas, including cyberterrorism. One online forum popular with supporters of terrorism (called Qalah, or "Fortress"), has a discussion area called "electronic jihad" in which potential al-Qaeda recruits can find links to the latest computer-hacking techniques. Iman Samudra, who was convicted and sentenced to death for taking part in the 2002 bombings of two Bali nightclubs, wrote a book titled *Aku Melawan Teroris* (I Fight the Terrorists). In this 2004 book, Samudra advocated that Muslim youth actively develop hacking skills "to attack U.S. computer networks." Samudra names several websites and chat rooms as sources for increasing hacking skills (Rollins and Wilson 2007, 15).

The real threat appears to come from hackers linked with state sponsors of terrorism. In April 2004, the US Department of State listed seven designated state sponsors of terrorism: Cuba, Iran, Iraq, Libya, North Korea,

Syria, and Sudan (Perl 2004).[7] Some of these countries may be involved in sponsoring and promoting cyberterrorism, and others already are. There are cyberterrorists linked with Iran, Iraq, North Korea, and Syria. Today, Iran is one of the world's most notorious sponsors of terror groups like Hezbollah, Hamas, Palestinian Islamic Jihad, the al-Aqsa Martyrs' Brigades, and various militant groups in Iraq. As cyberterrorist efforts begin to look more fruitful, Iran is working to develop the virtual capacities of its proxies. This currently means sending computer and network equipment, security packages, and relevant software, but it also could mean in-person training of cyberterrorists in Iran or by skilled Iranian cyberteams.

The Syrian government, for example, is certainly behind a series of cyberattacks launched by the Syrian Electronic Army hackers' group. According to the group's website, the Syrian Electronic Army was created in 2011 "when the Arab media and Western [media] started bias in favor of terrorist groups that have killed civilians [and] the Syrian Arab Army, and [destroyed] private and public property" (Syrian Electronic Army 2014). The attacks it has carried out since the start of its operation demonstrates that its mission is to advance Syrian interests through the use of cyberattacks. It is not involved in protecting Syrian websites or computer systems, but rather chooses to execute attacks against those it considers to be domestic and foreign enemies of Syria. The group's various activities attest to its central targets: government officials in countries throughout the region, Western and Arab media outlets, and recently even Internet media applications. In addition to its hacks on the Associated Press Twitter account and the CNN, *Time*, and *Washington Post* websites, in 2013 the Syrian Electronic Army hacked numerous high-profile social media accounts and websites associated with major news and human rights organizations: on January 9, the Saudi Defense Ministry and other Saudi government websites; on March 17, Human Rights Watch; on March 21, the BBC; on April 16, NPR; on April 21, CBS; on April 23, AP, as mentioned above; on April 29, the *Guardian*; on May 17, *Financial Times*; on May 26, Sky's Android Apps and Twitter account; and in July, @Thomsonreuters, Truecaller, Tango, and Viber, and the Twitter accounts of several White House staffers (Stalinsky and Sosnow 2013b).

---

[7] In light of recent political changes, some of these states have been removed from the State Department's list. Iraq was removed in 2004 following the 2003 US invasion, Libya was removed in 2006 after it decommissioned its nuclear program, and North Korea was removed in 2008 after it met nuclear inspection requirements. Cuba, Iran, Syria, and Sudan remain on this list as of September 2014 (Department of State 2014b).

Though there is no explicit known connection between the Syrian Electronic Army and the Syrian regime, the regime is believed to be behind the group's activities and has recognized its legitimacy. The Syrian Computer Society, which was headed by Bashar al-Assad before he became president in 2000 (and which is Syria's domain registration authority), is also believed to be connected to Syria's state security apparatus. In a June 2011 speech at Damascus University, President Assad compared online warriors to his military: "The army consists of the brothers of every Syrian citizen. . . . Young people have an important role to play at this stage, because they have proven themselves to be an active power." He added, "There is the electronic army, which has been a real army in virtual reality." ("Syria: Speech by Bashar al-Assad" 2011).

The link between terrorists and hackers presents another alarming scenario. As Eric Schmidt and Jared Cohen (2013, 39) predict regarding the rise of terrorist-hackers,

> Sudden access to technology does not in and of itself enable radicalized individuals to become cyber terrorists. There is a technical skills barrier that, to date, has forestalled an explosion of terrorist-hackers. But we anticipate that this barrier will become less significant as the spread of connectivity and low-cost devices reaches remote places. . . . Hackers in developed countries are typically self-taught, and because we can assume that the distribution of young people with technical aptitude is equivalent everywhere, this means that with time and connectivity, potential hackers will acquire the necessary information to hone their skills. One outcome will be an emergent class of virtual soldiers ripe for recruitment. Whereas today we hear of middle-class Muslims living in Europe going to Afghanistan for terror-camp training, we may see the reverse in the future. Afghans and Pakistanis will go to Europe to learn how to be cyber terrorists.

Terrorist groups, as well as governments and security agencies, are trying to recruit cybersavvy specialists and hackers to fight for their side. Recognizing how a cadre of technically skilled programmers enhances their destructive capacities, terrorists will increasingly target engineers, students, programmers, and computer scientists at universities and companies, building out the next generation of cyberwarriors. Such attempts have been recorded in the past and are only becoming more serious. In April 2012, FBI director Robert S. Mueller warned that "[t]errorists have shown interest in pursuing hacking skills and they may seek to train their own recruits or hire outsiders, with an eye toward pursuing cyberattacks. These adaptations of the terrorist threat make the F.B.I.'s counterterrorism mission that much more difficult and challenging" (quoted in Schmidt 2012). Indeed, it is hard to persuade someone to become a cyberterrorist, given the legal consequences, but money, ideology, religion, and blackmail will continue to play a large role in the recruitment process. As Schmidt and Cohen (2013, 162) noted, "Unlike governments, terrorist groups can play the antiestablishment card, which may strengthen their case among some young and disaffected hacker types. Of course, the decision to become a cyber-terrorist is almost always less consequential to one's personal health than signing up for suicide martyrdom."

Finally, the most threatening combination may be the emerging combination of state-terrorists-hackers. This triangle is not just viable, but in fact is already functional. Iran, for example, has produced a number of well-known hacker groups since the Internet was first introduced to the public in 2000 (Wheeler 2013). Today, Iran's Cyber Force is one of the most powerful in the world. Their force of cyberwarriors has executed a number of crippling attacks, and their cyberwarriors are among an extensive and secretive network of hackers, some of which cannot be traced to any one particular group. The majority of attacks stemming from Iran are DDoS attacks targeted at US banking institutions, including Wells Fargo, Bank of America, PNC, and Citigroup Citibank. The Iranian Revolutionary Guard first proposed the establishment of the Iranian Cyber Army (ICA) in 2005, but its implementation was accelerated as media attacks against the Ahmadinejad administration grew. Because of the ICA's development, Iran's cyber capabilities increased dramatically in a short number of years. ICA began to make its presence known in late 2009, after the Stuxnet virus attack on Iranian nuclear facilities. In response to these attacks, Iranian officials focused on developing cyberdefensive measures, but also have explored more offensive measures. In 2009, the American security and military institute Defense Tech included Iran among the top five in its list of the most powerful countries in terms of cyber force. Defense Tech also stated that the ICA was a subdivision of Iran's Revolutionary Guard cyber team, with an annual budget of $76 million and more than a billion-dollar investment in infrastructure. The ICA also enjoys access to large pool of talented hackers (Wheeler 2013).

Since its creation, the ICA has launched numerous cyberattacks. In December 2009, the ICA attacked Twitter, making it inaccessible in some countries and redirecting the users to an English-language webpage, which contained the following message: "This site has been hacked by the Iranian Cyber Army. . . . The USA thinks they control and manage Internet access, but they don't. We control and manage the internet with our power . . ." (Beaumont 2009). In February 2011, the ICA attacked the Voice of America's website, replacing its Internet home page with a banner bearing an Iranian flag and an image of an AK-47 assault rifle. It left a message on the Voice of America sites that stated, "We have proven that we can." It also called on the United States to stop interfering in Islamic countries. Later, an Iranian government official announced that the Iranian Revolutionary Guard Corps was behind a recent computer attack that disrupted Voice of America Internet programming (Gertz 2011). In 2011, the ICA reportedly hacked into 500 Internet security certificates and then used them to attack around 300,000 Iranian Internet users. According to the Dutch government, attackers stole the certificates from DigiNotar, a Dutch web security firm. The Dutch Justice Ministry published a list of the users of fake certificates that were sent to sites operated by Yahoo!, Facebook, Microsoft, Skype, AOL, the Tor Project, WordPress, and intelligence agencies (Associated Press 2011).

On June 2013, Israeli prime minister Benjamin Netanyahu told a conference on cyberwarfare that Israel's computer systems are subject to nonstop cyberattacks from Iran. He claimed that critical infrastructure, including that in the power, water, and banking sectors, have all come under cyberattack. He added, "In the past few months, we have identified a significant increase in the scope of cyber-attacks on Israel by Iran. These attacks are carried out directly by Iran and through its proxies, Hamas and Hezbollah" (quoted in Heller 2013).

In an escalation of Iranian cyberintrusions targeting the US military and critical infrastructure, in September 2013, US officials reported that Iran hacked unclassified Navy computers. The officials said that the attacks were carried out by hackers working for Iran's government or by a group acting with the approval of Iranian leaders. Later, in February 2014, it was reported that the attacks were more extensive than first thought: the cyberattack targeted the Navy Marine Corps Internet, which is used by the Navy Department to host websites; store nonsensitive information; and handle voice, video, and data communications. The *Wall Street Journal* reported that the hackers were able to remain in the network until November 2013. Thus, it took the Navy about four months to finally purge the hackers from its biggest unclassified computer network (Gorman and Barnes 2014). Also in 2013, it was reported that Iranian hackers were able to gain access to control-system software that could allow them to manipulate American oil or gas pipelines (Kumar 2012). Control systems run the operations of critical infrastructure, regulating the flow of oil and gas or electricity, turning systems on and off, and controlling key functions. The hacking campaign, which the United States believes has direct backing from the Iranian government, focused on the control systems that run oil and gas companies and, more recently, power companies, current and former officials said (Gorman and Yadron 2013).

The threat of cyberterrorism is certainly alarming and dreadful. However, in the virtual war between terrorists and counterterrorism forces and agencies, the actions that the terrorists themselves have taken can suggest possible countermeasures. These countermeasures, both technological and psychological, are the subject of the following chapters.

Chapter 9

Countermeasures:
"Noise" and the M.U.D. Model

[I]t is essential to appreciate the very strong collective ethos of
the Internet. From its inception, Internet users have always been
passionately in favour of internal control and against outside
influence. In effect, for many years the Internet has operated as a fully
functioning anarchy.

—Duncan Langford, "Ethics @ the Internet:
Bilateral Procedures in Electronic Communication" (1998, 98)

The Challenge: Counterterrorism Online

The Internet and its online platforms provide terrorists with anonymity,
low barriers to publication, and low costs of publishing and managing content. The advent of greater user interactivity provided by Web 2.0 and the

This chapter expands on material originally published in "Applying the Notion of Noise to Countering Online Terrorism" (Von Knop and Weimann 2008).

meteoric rise of social media have enabled radical groups and terrorists to freely disseminate ideas and opinions using such multiple modalities as websites, blogs, forums, and social networking and video-sharing websites. Counterterrorism on the Internet is certainly lingering behind the terrorists' manipulative use of this medium. Given the growth of Internet research in recent years, it is rather surprising that research into online countermeasures has been overlooked, or at least has not provided efficient strategy and fruitful devices or tactics. According to Jody Westby (2006), several factors combine to explain this research gap: (1) difficulties in tracking and tracing cyber communications; (2) the lack of globally accepted processes and procedures for the investigation of cybercrimes and cyberterrorism; and (3) inadequate or ineffective information sharing systems between the public and private sectors, between governments, and between counterterrorism agencies. Although there are technological reasons that increase the difficulty of tracking and monitoring online terrorist traffic, the problem rests largely with the legal domain: governments around the globe must address these critical issues through legal frameworks and policy directives that advance information security and improve the detection and prosecution of cyberterrorists.

The virtual war between terrorists and counterterrorism forces and agencies is certainly vital, dynamic, and ferocious. The National Security Agency, Department of Defense, Central Intelligence Agency, Federal Bureau of Investigation, Defense Intelligence Agency, other US and foreign intelligence agencies, and some private contractors have been fighting back: cracking terrorist passwords, monitoring suspicious websites (and cyberattacking others), and planting bogus information. Interest in countering online terrorism has also brought together researchers from around the world and from various disciplines, including psychology, security, communications, and computer sciences, to develop tools and techniques to respond to the challenge (Sinai 2011). It has spawned an interdisciplinary research area—intelligence and security informatics—which studies the development and use of advanced information technologies and systems for national, international, and societal security-related applications. However, as some security and terrorism experts such as Bruce Hoffman argue, there could be better ways to counter the threat: "The government efforts are inadequate. The private sector is doing a better job than the government. Our enemies have embraced the Internet. We have to ask how closely the government is monitoring it" (quoted in Blumenthal 2007).

Recognizing the online threat, the White House's counter-radicalization strategy, published in August 2011, acknowledged "the important role the

Internet and social networking sites play in advancing violent extremist narratives" (The White House 2011a, 6). The strategy's implementation plan, which came out in December 2011, stated that "the Internet has become an increasingly potent element in radicalization to violence" and that new "programs and initiatives" had to be "mindful of the online nature of the threat" (The White House 2011b, 20). Crucially, it also committed the administration to formulate a strategy in its own right: "[B]ecause of the importance of the digital environment, we will develop a separate, more comprehensive strategy for countering and preventing violent extremist online radicalization and leveraging technology to empower community resilience" (The White House 2011b, 20).

However, no such online strategy has yet been published. The Bipartisan Policy Center's Homeland Security Project, co-chaired by former 9/11 Commissioners Gov. Tom Kean (R-NJ) and Rep. Lee Hamilton (D-IN), released a report in December 2012, *Countering Online Radicalization in America*, that identified the shortcomings in US online counter-radicalization strategy and recommended a path to improvement (Neumann 2012). According to the center's analysis, approaches aimed at restricting freedom of speech and removing content from the Internet are not only the least desirable strategies, but they are also the least effective. Instead, the government should play a more energetic role in reducing the demand for radicalization and violent extremist messages. Specifically, the center's report recommended developing a strategy:

- The White House must revise its counter-radicalization strategy in order to make it stronger and more specific; and the White House should begin its implementation with alacrity.
- The strategy should include components designed to reduce the demand for radicalization and violent extremist messages, and exploit the online communications of extremists in order to gain intelligence and gather evidence.
- The government should clarify online law enforcement authorities, and communicate with Internet companies on the nature of radical threats, propaganda, and communication.
- The government should accelerate the establishment of informal partnerships to assist large Internet companies in understanding national security threats as well as trends and patterns in terrorist communications, so that companies become more conscious of emerging threats and key individuals and organizations, and may find it easier to align their takedown efforts with national security priorities.

- The government should work to reduce the demand for radical messages by encouraging civic challenges to extremist narratives through countermessaging efforts by community groups to promote youth awareness and education.
- Counterextremism education for youth should be expanded, along with partnerships to educate parents, teachers, and communities on the signs and risks of extremism. For example, government should encourage school authorities to review and update their curricula on media literacy, consider violent extremism as part of their instruction on child-safety issues, and develop relevant training resources for teachers.
- The government should identify up front what resources will be committed to outreach and education programs, as well as the metrics that will be used to measure success.
- Law enforcement and intelligence agencies need to take better advantage of the Internet to gather intelligence about terrorists' intentions, networks, plots and operations; and they need to secure evidence that can be used in prosecutions.
- The amount of online training offered to members of law enforcement and intelligence agencies should be increased, including state and local agencies, so they are conscious of the increasingly virtual nature of the threat and can use online resources to gather information about violent extremist communities in their local areas. For example, extremist forums and social networking sites are essential for identifying lone actors, many of whom have a long history of online activism through posting messages in online forums, running blogs, and maintaining Facebook pages. These communications should be monitored to watch for sudden changes in behavior including escalating threats or announcements of specific actions. (Neumann 2012)

Although this list of recommendations may be useful and effective, it does not represent a general strategy or theory. Besides the legal and practical issues involved, online counterterrorism efforts suffer from a lack of strategic thinking. Various measures have been suggested, applied, replaced, changed, and debated, but there has never been an attempt to propose a general model of online counterterrorism strategy. Countering terrorist use of the Internet to further ideological agendas will require a strategic, government-wide (interagency) approach to designing and implementing policies to win the war of ideas. The notion of "noise" in communication theory is suitable as a basic theoretical framework to conceptualize various measures and their applicability.

Noise in Communication Processes

In communication theory, noise is that which distorts the signal on its way from transmitter to recipient. Noise interferes with the communication process, as it keeps the message from being understood and prevents it from achieving its desired effect. It is inevitable that noise distorts the message being sent by getting in the way. The concept of noise was introduced as a concept in communication theory by Claude Shannon and Warren Weaver in the 1940s.[1] They were mostly concerned with mechanical noise, such as the distortion of a voice on the telephone or interference with a television signal that produced "snow" on a television screen. In the succeeding decades, other kinds of noise have been recognized as potentially important problems for communication (Rothwell 2004):

- *Physical noise* is any external or environmental stimulus that distracts us from receiving the intended message sent by a communicator.
- *Semantic noise* occurs because of the ambiguities inherent in all languages and other sign systems.
- *Cultural noise* occurs when the culture or subculture of the audience is so different from that of the sender that the message is understood in a way that the sender might not have anticipated.
- *Psychological noise* results from preconceived notions we bring to the communication process, such as racial stereotypes, reputations, biases, and assumptions.

Although the concept of noise was first perceived as relevant only to the potential for interfering with the transmission of a message, it later became recognized as a crucial element in the communication process, potentially affecting each stage of the process. Noise can represent physical noise (e.g., loud music, shouting reporters) as well as mental noise such as stress, anxiety, or time constraints that distracts one's thoughts. Today, noise is used as a metaphor for all the problems associated with effective communication,

---

[1] Their final model was published in *The Mathematical Theory of Communication* (Weaver and Shannon 1963).

interfering with every stage of the process (including encoding, decoding, transmitting, and interpreting).

The concept of noise in communication theory and research often has been treated as a negative element, damaging the communication process. In fact, most empirical uses were directed at reducing or minimizing noises to improve the flow of communication. However, today noise is breaking away from the status of undesirable phenomenon bestowed upon it by traditional communications theory. No longer merely an undesirable element to be eradicated so as to retain the purity of the original signal, noise can be regarded as a more complex, even desired element. When considering terrorist (or any other illegal, harming, or dangerous) communication, one may question the instrumentality of creating noise that may reduce the communicator's efficiency and success. As the following sections illustrate, noise can be employed to harm the flow of information, the decoding of messages, the communicator's credibility and reputation, the signal's clarity, the channel's reach, the receivers' trust, and other components of terrorist communication. Semantic, psychological, cultural, and physical noises may all be part of a rich variety of countermeasures, which can be organized in a strategic framework. Thus, noise could become a key conceptual and theoretical foundation in the strategy of countering terrorism online (Aly, Weimann-Saks, and Weimann 2014).

Conceptualizing the Notion of Noise into a Strategy

Strategic communication planning is one of the most neglected areas of counterterrorism, especially when it comes to the disruption of terrorist communication (Bolz, Dudonis, and Schulz 2002; Halloran 2007). Strategic communication requires a sophisticated method that maps perceptions and influences networks, identifies policy priorities, formulates objectives, focuses on "doable tasks," develops themes and messages, employs relevant channels, leverages new strategic and tactical dynamics, and monitors success. This approach has to build on in-depth knowledge of radical thinking, radicalization processes, and factors that motivate radical or terrorist behavior. A successful approach will have to combine hard and soft power in terms of strategic communication (Nye 2004a). The principal understanding of power—to make others do what you want or produce the outcomes you want—has not changed. Although hard power is the ability to order others to do what you want, Joseph Nye (2004b) defines soft power as "the ability

to get what you want through attraction rather than coercion or payments."
As Nye (2004b, 256) explains: "When you can get others to want what you
want, you do not have to spend as much on sticks and carrots to move them
in your direction. Hard power, the ability to coerce, grows out of a country's
military and economic might. Soft power arises from the attractiveness of
a country's culture, political ideals, and policies."

An effective communication strategy to counter online terrorist activities has to combine both hard power elements (such as hacking) and soft
power elements (such as psychological warfare), because "soft power and
hard power can reinforce each other; one is not contrary to each other"
(Nye 2003, 46). Nye (2003) also assumes that soft power does not increase
relative power on the hard side, but it does make hard power more acceptable, lowering the costs of exercising such power. Returning to the notion
of noise, mechanical noise includes the elements of "hard power," whereas
social and psychological noise both require a "soft power" approach. A successful application of "noise" as a counterstrategy against terrorists' appeal
on the Internet will need to incorporate the following elements:

- *Credibility:* As in other communication processes, a key factor in
  determining the persuasiveness of terrorist messages is the credibility of the source. Thus, a counterstrategy may involve systematically
  damaging the credibility of the terrorist authority while introducing
  an alternative authoritative or credible source. Previous studies of the
  inner debates and disputes among terrorists online may expose cleavages and splits that can be used to attack the credibility of online terrorist authorities (Weimann 2006c, 2009d).
- *Terminology:* Terminology plays a significant role in the process. For
  example, a better understanding of the nuances of Islam (in contrast
  to extreme Islamic preaching), the nuances of jihad, Salafist terminology, and the subtexts of their communication will be required. Effective application of "noise" may rely on the use of key terms, exposing
  their manipulative uses, relating new meaning to them, or weakening
  their conventional usage. For instance, should the terms "jihad" and
  "jihadist" be replaced with their nonviolent interpretations, provided
  by leading Muslim figures? The term "Islamism" may be replaced
  with the term "anti-Islam," suggesting that it is far from being in line
  with the pure and original values of Islam.
- *Traditions:* The rhetoric of online preaching and radicalization relies
  on traditions. Thus, solutions should come from within Muslim

traditions, because solutions derived from Western traditions will by definition be rejected as illegitimate. This approach requires a deep understanding of the traditional thinking, values, and symbols of the community targeted.

- *Partners:* To be truly effective in delegitimizing radical appeals and attraction, the countercampaign may involve the activation of "partners" who come from within the targeted community. Thus, the alternative voices, suggesting alternative narrative and discourse, should come from within Muslim religious leadership, not from the West.

- *"Think global, act local":* A long-term perspective starts with the search for additional "agents" of change in supporting actors and institutions. Such actors may include state institutions that provide education, medical treatment, and social warfare. Furthermore, universities in the Arab world (as well as schools and nongovernmental organizations) could provide useful venues in which to open up the "modernization" debate. The same debate can be continued through Islamic education in public schools throughout Europe, articles in journals, and the efforts of individual intellectuals. Such debate would allow all parties to hear different views and engage in a discussion that can truly educate, without appearing to be a narrowly construed propaganda effort from the start. Again, these educational reforms need to come from within the Muslim world, although the West can support it with its resources (Von Knop and Weimann 2008, 891–92).

An effective strategy must be multifaceted, addressing all of these aspects. It will be a long-term undertaking that must be based on familiarity with the targets' background, mentality, values, beliefs, history, frustrations, and hopes. Moreover, before such a communication strategy can be developed, it is essential to define the strategic goals, identify potential partners, and characterize the target audiences. As one US counterterrorism official explained, the problem is "that we focus on the terrorists and very little on how they are created. If you looked at all the resources of the US government, we spent 85–90 percent on current terrorists, not on how people are radicalized" (quoted in deYoung 2006).

The first step in this process involves identifying terrorists' online platforms and studying their contents to determine the necessity of applying various disruptive tactics. This monitoring can be done by human analysts and coders (i.e., the manual approach) or by automatic web crawlers. The manual approach is often used when the relevance and quality of information from websites are of the utmost importance; however, it is labor-intensive and time-consuming, and often leads to inconclusive results. The automatic web-crawling technique is an efficient way to collect large amounts of web pages. Crosslingual information retrieval can help break language barriers by allowing users to retrieve documents in foreign languages through queries in their native languages.

The monitoring process will have to cover both websites and social media, from forums and chatrooms to Facebook, Twitter, Youtube, and Instagram. Observation is essential to learn about the target groups, participants, key players, appeals and rhetorical motives, and ideas and rewards promised. The information gathered at this stage may indicate what measures are required, if any, and the urgency for applying noise tactics. The following illustrative examples of actual "noise" will distinguish between mechanical and technological noises and psychological and social noises.

Applying Noise

Mechanical and Technological Noises

These types of noise refer to the technological disruption of the flow of communication. Mechanical and technological tactics include a rich variety of interventions, from damaging websites and defacing and redirecting their users to spreading viruses and worms, blocking access, hacking, and total destruction. These deviant measures can be adopted and used against online terror in order to minimize its reach and impact. In the most severe cases, hacking the websites may be the most extreme measure, though in the long run it is not always the most efficient one. Ross Anderson (2008) describes an impressive arsenal of common hacking techniques, all of which counterterrorism strategies can employ. Many actual attacks, Anderson argues, involve combinations of vulnerabilities. Anderson's list of "the top 10 vulnerabilities" suggests a hacking strategy using each of these vulnerabilities (2008, 368). Most of the exploits make use of program bugs, of which the majority is stack overflow vulnerabilities. (These vulnerabilities are flaws in program coding that can make a program crash if too much data is forced onto it, as a hacker might attempt to do.) Anderson also argues that none of these attacks can be stopped by encryption, and not all of them are stopped by firewalls.

Such disruptive counterattacks on terrorist online platforms are not new. In May 2012, Secretary of State Hillary Clinton revealed that the US government had intruded on al-Qaeda websites in an effort to counter the terrorist group's activities (Hudson 2012). This revelation showed not only that the fight against terror continues unabated, but also that much of the battle is taking place online. According to Secretary Clinton, State Department cyber-experts targeted tribal websites in Yemen. The intruders took down the ads that al-Qaeda had put up on these sites, which bragged about killing Americans, and uploaded their own counter-ads, which exposed the ruthless coercive methods that al-Qaeda has used on Yemenis. Numerous states, including France, Germany, Israel, Italy, the Netherlands, Russia, and the United Kingdom, have launched similar destructive and disruptive attacks. However, such attacks have had very limited effect, since the terrorists have easily managed to reestablish their online platforms and reemerge in cyberspace.

A more sophisticated form of "mechanical noise" is the optional use of Trojan horses, viruses, and worms against terrorists. The common distinction among the three is that a Trojan horse is a program that does something malicious when run by an unsuspecting user; a worm is a program that replicates itself in order to spread to other computers; and a virus is a worm that replicates by attaching itself to other programs (Anderson 2008). A virus or a worm will typically have two components: a replication mechanism and a payload. The replication mechanism enables the virus or worm to make a copy of itself somewhere else, usually by breaking into another system or by mailing itself as an attachment. The payload—which usually is activated by a trigger, such as a specific date—may then inflict one or more of these damages: make selective or random changes to the computer's protection; make selective or random changes to user data (e.g., trash the disk); lock the network (e.g., by replicating at maximum speed); steal resources or steal data; and even take over the infected system.

One further form of this type of noise is identity theft. Terrorists use this crime to fund their activities and disguise the identities of their operatives. For example, the al-Qaeda terrorists involved in the 9/11 attacks had opened 14 bank accounts using several different names, all of which were fake or stolen. If criminals can steal someone's identity, counterterrorists can do the same. Using identity theft tactics, a video or audio recording could be produced and placed on al-Qaeda websites. Such attacks can create confusion among the terrorists' followers and supporters, harm the credibility of their websites and messages, and lower these sites' exposure and attraction.

Such actions require the skills and experience of hackers. Indeed, hackers have been secretly recruited and used for counterterrorist attacks. In September 2013, the British government officially admitted such practices:

it announced the establishment of the Joint Cyber Reserve Unit. Under the £500 million initiative, the British Ministry of Defence would recruit hundreds of reservists as computer experts to work alongside regular armed forces. The unit was formed to defend national security by safeguarding computer networks and vital data, and also to launch strikes in cyberspace if necessary. The head of the new unit declared that convicted computer hackers could be recruited to the special force if they pass security vetting (BBC 2013). Many companies and security agencies are also recruiting hackers, either to play the "red team" in simulations of cyberattacks, or to suggest and launch counterattacks. The well-regarded Mandiant Corporation, which uncovered a series of cyberattacks on US networks by a branch of China's People's Liberation Army, was also hired by the *New York Times* and the *Wall Street Journal* when they were hacked. Mandiant's professional hackers also consult with a number of Fortune 500 companies at a reported rate of $450 an hour (Mulrine 2013). The US military is also reaching out to even younger students through high school talent searches in the form of cybergames like CyberPatriot, a hacking tournament that pits young high school students against industry mentors (who play the aggressors) in a contest to see who can destroy the other's network first (Mulrine 2013).

In the new cyberspace battlefield, there is a real need for new warriors: the cyberwarriors. As John Arquilla, a professor of defense analysis at the US Naval Postgraduate School in Monterey, and the man who coined the term "cyberwarfare" argued,

> Instead of prosecuting elite computer hackers, the US government should recruit them to launch cyber-attacks against Islamist terrorists and other foes. The brilliance of hacking experts could be put to use on behalf of the US in the same way as German rocket scientists were enlisted after the second world war . . . the US had fallen behind in the cyber race and needed to set up a "new Bletchley Park" of computer whizzes and code-crackers to detect, track, and disrupt enemy networks. If this was being done, the war on terror would be over. (Quoted in Carroll 2012)

Applying Psychological and Social Noises

The tactics included in this category involve various psychological and social operations and counterpropaganda. Psychological and social noises encompass several different terms: information warfare, information operations (IOs), and psychological operation (PSYOP). There are numerous definitions of these terms (Denning 1998; Ventre 2009, 2011). Information warfare is the use and management of information technology to pursue a competitive advantage over an opponent. It may involve collecting tactical information, ensuring that existing information is valid, spreading propaganda or disinformation to demoralize or manipulate the enemy and the public, or denying opposing forces the opportunity to collect information or undermining the quality of their information. Information warfare consists of a broad variety of IOs. The focus of IO is on the decision maker and the information environment in order to affect decisionmaking and thinking processes, knowledge, and situational understanding. Thus, for instance, PSYOPs are a part of IOs.

As defined, PSYOPs are operations designed "to convey selected information and indicators to foreign audiences to influence their emotions, motives, objective reasoning, and ultimately the behavior of foreign governments, organizations, groups, and individuals. The purpose . . . is to induce or reinforce foreign attitudes and behavior favorable to the originator's objectives" (Joint Chiefs of Staff 2010, G-8). As a communication medium and vehicle of influence, the Internet is a powerful tool for psychological campaigns. Consequently, the realm of military PSYOP must be expanded to include the Internet: "Although current international law restricts many aspects of PSYOP either through ambiguity or noncurrency, there is ample legal room for both the U.S. and others to conduct PSYOP using modern technology and media such as the Internet" (Lungu 2001, 17). Whether used offensively or defensively, it is clear that the Internet is an important tool for PSYOP and can bring tremendous capabilities and informational advantage to forces employing this medium. Given the strategic opportunities afforded by the Internet, there are several options for employing this medium in PSYOP. Counterterrorism agencies in particular could use the Internet offensively to help achieve unconventional warfare objectives, as well as to address and counter adversarial propaganda, disinformation, and incitement. In addition to developing websites with this purpose in mind, preemptive messages and Internet products such as streaming audio and video, online video games, mediated newsgroups, and ad banners can be leveraged for their strategic value and reach. A 2000 Defense Science Board report on PSYOP also suggested some less obvious potential tools that use emerging media technologies, such as chat rooms and instant messaging services that could be used for "guided discussions" to influence how various groups and audiences think about certain topics (quoted in Lungu 2001, 15–16).

Some of these tactics are rooted in the evolving domain of political Internet campaigns. MoveOn.org is a prominent example of a web-based political campaign, in the form of a public policy action group. In many ways, terrorists launch their online campaigns in the same way legitimate political campaigns use the Internet. Both types of campaign attempt to attract and seduce users by engaging them in a sensory experience, trying to manipulate their needs, suggesting the fulfilment of a goal, and providing a higher-level motivation to inspire and guide them to make a choice. Once the goal is fulfilled and the user is captivated, the function at this point is to form a relational bond between the user and the party, candidate, group, or organization. Online campaigns and countercampaigns in the political arena can therefore provide lessons to be learned and serve as pivotal experiments to guide counterterrorist campaigns.

Political campaigns are in fact a series of actions and appeals involving resource mobilization. The communicators are trying to mobilize the predisposed, demobilize hostile voters, convince the undecided, and convert the initially hostile. They must do so by designing persuasive messages, communicating these messages, monitoring the responses, and facilitating the desired behavior. Campaigning via the interactive Internet often provides social bonding and replicates feelings of personal contact. These elements, which are also frequently found in terrorist websites, also can be used in countercampaigns. However, before such campaigns are launched, the agencies involved should know both the psychographic profiles of those susceptible to recruitment and the messages that affect them. Agencies also need to understand how these individuals are influenced: what channels are meaningful to them, who they listen to, how peer networks affect them, and how to reach them most effectively.

In a January 2014 *Los Angeles Times* op-ed on "Future Terrorists," Jane Harman of the Wilson Center argued that "we need to employ the best tools we know of to counter radicalizing messages and to build bridges to the vulnerable. . . . Narratives can inspire people to do terrible things, or to push back against those extremist voices" (Harman 2014). To run such a strategy, a political Internet campaign against terrorism must use tactics that have proven to be successful and that have counterterrorism applications. Finding such effective tactics was at the heart of discussions at the January 2011 Riyadh Conference on the Use of the Internet to Counter the Appeal of Extremist Violence. Co-hosted by the United Nations Counter-Terrorism Implementation Task Force and the Naif Arab University for Security Sciences in Riyadh in partnership with the Center on Global Counterterrorism Cooperation, the conference brought together around 150 policymakers, experts, and practitioners from the public sector, international organizations, industries, academia, and the media (United Nations Counter-Terrorism

Implementation Task Force 2012). The conference focused on identifying good practices in using the Internet to undermine the appeal of terrorism, expose its lack of legitimacy and its negative impact, and undermine the credibility of its messengers. Key themes included the importance of identifying the target audience, crafting effective messages, identifying credible messengers, and using appropriate media to reach vulnerable communities. Among the recommendations were several that relate to psychological/social noises:

- Promote counternarratives through all relevant media channels (online, print, television/radio).
- Make available a counternarrative whenever a new extremist message appears on Facebook, YouTube, or similar outlets.
- Offer rapid counternarratives to political developments (e.g., highlight the absence of al-Qaeda and other extremist groups at popular protests).
- Consider selective take-down of extremist narratives that have the elements of success.
- Ensure that counternarratives include messages of empathy/understanding of political and social conditions facing the target audience, rather than limiting the counternarrative to lecturing or retribution.
- Offer an opportunity for engagement in crafting and delivering counternarratives to young people who mirror the "Internet Brigade" members of al-Qaeda.
- Support the establishment of civil society networks of interested groups, such as women against violent extremism, parents against suicide bombers, or schools against extremism. (United Nations Counter-Terrorism Implementation Task Force 2012, 81)

The M.U.D. Model

This interference with terrorist online communications may be described by the "M.U.D." model. This model was presented first at the September 2006 NATO workshop "Hypermedia Seduction for Terrorist Recruiting" held in Eilat, Israel, and then at the October 2007 NATO advanced research workshop "Responses to Cyber Terrorism" held in Ankara, Turkey. The M.U.D. approach (which stands for monitoring, using, and disrupting) is a flexible, multistep model that applies the options of passive surveillance (monitoring), interfering with the traffic and the online contents (using), and finally removing material and blocking access (disrupting) (Sinai 2011, 23–24). Figure 9.1 presents a visual representation of the M.U.D. model.

First, terrorist websites need to be monitored in order to learn about terrorist mindsets, motives, persuasive "buzzwords," audiences, operational plans, and potential targets for attack. This form of knowledge discovery refers to nontrivial extraction of implicit, previously unknown and potentially useful knowledge from data. Increasingly, forums, blogs, and other frequently updated sites have become the focus of monitoring attention. Second, counterterrorism organizations need to "use" the terrorist websites

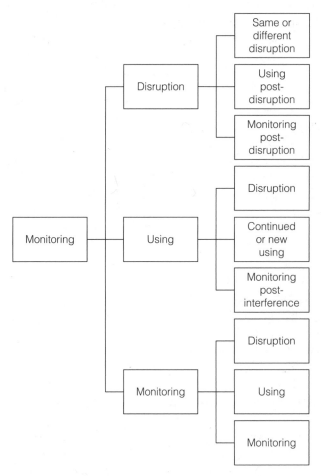

Figure 9.1.   The M.U.D. Model

to identify and locate their propagandists, chatroom discussion moderators, Internet service provider hosts, operatives, and participating members. The retrieved data need to be archived to enhance the learning process and to identify social networks. A social network consists of a network of connections between people, between people and events, and between people and organizations. Mathematical techniques can be used to identify key persons or clusters of people within a network, optimize the way in which a network is displayed, and measure the network's robustness. Third, terrorist websites and other online platforms such as social media need to be "disrupted" through negative and positive means. In a negative "influence" campaign, sites and postings can be infected with viruses and worms to destroy them, or kept "alive" in order to be flooded with false technical information about weapons systems, rumors intended to create doubt about the reputation and credibility of terrorist leaders, or conflicting messages that can be inserted into discussion forums to confuse operatives and their supporters. In a more positive approach, alternative narratives can be crafted and inserted into the sites to demonstrate the negative results of terrorism or, if targeting potential suicide bombers, to suggest the benefits of the "value of life" versus the self-destructiveness of the "culture of death and martyrdom" (Sinai 2006).

These options are not mutually exclusive. The process of monitoring online contents can lead to a stage of interference that may, in turn, change to a disruptive stage. The model above describes these variations. The concept of "noise" is well integrated into the M.U.D. model, since the various noises (mechanical, psychological, social, and institutional) may be applied in each stage, in varying forms and magnitudes. Various producers present, post, and promote radical message in various forms (videos, lectures, games, postings, online publications, and social media) on the Internet and in the dark web. Target audiences such as potential followers, radicals, terrorist organizations, journalists, governmental agencies, and nongovernmental organizations receive these messages directly or indirectly, through exposure, interpersonal diffusion, or search engines. Greater understanding of this communication process enables counterterrorism proponents to identify potential targets of a counterstrategy or a type of noise. The model in figure 9.2 presents the terrorist communication process and the placement of various "noises" that may hinder, slow down, damage, or disrupt terrorist abuse on the Internet (Von Knop and Weimann 2008).

No longer merely an undesirable element to be eliminated from communication, noise can be regarded as a more complex and even desired element under certain circumstances. When it comes to terrorist communication, the concept of noise can serve as a key conceptual and theoretical foundation in the strategy of countering terrorism online. As this model demonstrates, various "noises" are useful methods of harming the flow, the decoding, the communicator's credibility and reputation, the signal's clarity, the channel's reach, the receivers' trust, and more aspects of terrorist messages online. By creating and using mechanical/technological or social/psychological noises, counterterrorism proponents may tap into the potential of a rich variety of countermeasures and help to organize them in a strategic framework.

The notion of noise also relies on using the vulnerabilities of terrorist online activities. Terrorists' online presence and activities are mostly

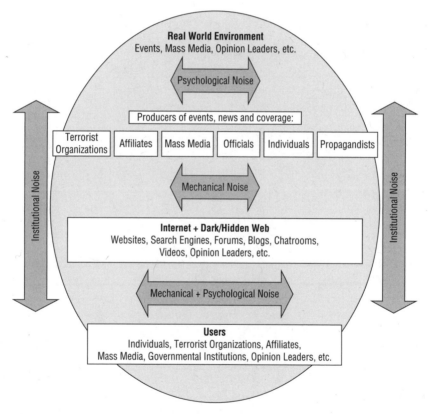

Figure 9.2.   Using "Noises" for Counterterrorist Online Communication

Source: Von Knop and Weimann 2008, 893.

"visible"; they are open to all. Terrorists are seeking greater exposure and trying to reach vast audiences, particularly when it comes to propaganda, psychological warfare, publicity, and the early stages of the radicalization process. Moreover, the way they invade the Internet and abuse its liberal spirit and unregulated nature leaves them open to intrusions. Such intrusions may and should include efforts to challenge the seductive terrorist narrative with alternative counternarratives.

Chapter 10

The War of Narratives

More than an armed confrontation, the war on terrorism is being played in the realm of narratives and it involves ideas, values, and images. The studies on terrorist online propaganda and radicalization identify the terrorist narratives being strategically deployed by al-Qaeda and its affiliates, jihadists, and other militant groups. These narratives are used to fuel extremism and attract new recruits. To develop a strategy and to identify appropriate tactics to counter terrorists' narratives, it is necessary to gain a deeper understanding of the role that these narratives play in seducing and persuading target audiences. Thus, it appears that the most effective way for Western democracies to counter terrorism is to monitor the emerging terrorist narratives and launch credible counternarratives (Archetti 2010). This chapter presents the complexity of terrorist online narratives and their newest versions. These narratives are drawn from terrorist websites, blogs, videos, Facebook groups, Tweets, and other types of online media. It will then examine the use of strategic counternarratives in terms of targeted audiences, contents, and visions.

Seductive Narratives

There can be no compelling counter-narrative until the extremist
narrative itself is well understood.

—Homeland Security Policy Institute/
Critical Incident Analysis Group (HSPI/CIAG),
NETworked Radicalization: A Counter-Strategy (2007, 15)

Narratives are essentially "compelling storylines which can explain events
convincingly and from which inferences can be drawn" (Freedman 2006,
22). A narrative is a story, and the word itself derives from the Latin verb
narrare, "to tell." However, Lawrence Freedman (2006, 22) argues that narratives are "strategic stories" because they are "designed or nurtured with the
intention of structuring the responses of others to developing events." Online
platforms for storytelling are continuously expanding, as activists and terrorists alike recognize the persuasive power of narratives. What makes the
narrative form so compelling? Within a persuasive narrative, what aspects
motivate an audience to action? Numerous studies on narratives and persuasion reveal that narratives are persuasive when they touch our emotions,
relate to what we believe, teach us new behaviors, and work on our social
or cultural identity (Dal Cin, Zanna, and Fong 2004). Why is narrative persuasion especially effective?

> There are two general means by which narratives might overcome resis-
> tance, each of which reflects a variety of specific processes. First, narra-
> tives may overcome resistance by reducing the amount and effectiveness
> of counterarguing or logical consideration of the message. Second, nar-
> ratives may overcome resistance by increasing identification with char-
> acters in the story. (Dal Cin, Zanna, and Fong 2004, 177)

Narratives differ from rhetoric in the way that messages are communicated:
Whereas the aim of advocacy is to present rational, clear, logical, specific
arguments, the aim of a narrative is to tell a story. In a narrative, beliefs
are often implied as opposed to stated explicitly. This approach may inhibit
counterarguments because it leaves the reader with no specific arguments
to refute.

Narratives are often concerned with relating the life experiences of other
people, be they real or fictional. As Michael Slater (2002) has suggested, it
may be especially difficult to counterargue the lived experiences of another

real or fictional person. Although one might be able to argue against hypo-
thetical examples ("That would never happen"), it is much more difficult to
argue against another's "real" experiences as conveyed in a narrative. These
seductive narratives have "absorption" potential: researchers have highlighted the concept of absorption into a narrative (e.g., Gerrig 1993; Green
and Brock 2000; Green, Strange, and Brock 2002) and have demonstrated
that the cognitive and emotional demands of absorption into a narrative leave
readers with little ability or motivation to generate resistance (e.g., Green
and Brock 2000; Slater 2002). Absorption into a narrative is believed to be
a convergent process, where all mental faculties are engaged in the narrative experience (Green and Brock 2000). A reader who is absorbed into a
narrative loses access to real-world facts and suspends disbelief and doubts.
Melanie Green and Timothy Brock (2000, 701) termed this phenomenon
transportation—"a convergent process, where all mental systems and capacities become focused on events in the narrative." This transportation into
the narrative in turn causes narrative persuasion. Numerous studies, including those referenced above, support the idea that transportation is a general
mechanism that underlies persuasion through narratives. Green and Brock
(2000) also found that the more transported participants were, the more they
tended to endorse beliefs implied by the narrative.

The seductive narrative relies on personal involvement, or transportation, in response to the story. The emotional element in seductive narratives
plays a crucial role. The emotional appeal of liking and evoking positive
sentiments is meant to create identification. Identification, however, requires
likeness (perceived similarity) to a character, or some desire to be *like* the
character (Oatley 2002, Slater 2002). Narratives evoke feelings or emotions by allowing a person to identify with a character in a similar situation. Even if the narrative is exaggerated or abstract, the listener or reader
understands that he or she is not the first person to have undergone a particular scenario. The narrative reassures listeners or readers that their feelings of frustration, solitude, stress, anxiety, doubt, and other emotions are
not unique, and that others in their situation have felt the same emotions.
Studies have shown that narratives influence beliefs and behavior "in part
by encouraging empathetic and emotional connections with story characters" (Mazzocco and Green 2011, 28).

The power of narratives stems also from their unthreatening nature. Mark
Zanna (1993) has argued that resistance to persuasion should result when
the targeted individuals are faced with arguments that support an attitudinal
position that falls outside their latitude of acceptance. That is, people have

some degree of "wiggle room"—a latitude of acceptance—around their attitudes (see Sherif and Hovland 1961). The latitude of acceptance can vary in size, from very narrow (indicating a fairly rigid attitudinal position) to very wide (indicating a more flexible attitudinal position). Thus, it is necessary to avoid closed-mindedness on the part of the targeted individuals. Ideally, one should seek to present an argument that is fairly extreme without making listeners aware of its extremity (Dal Cin, Zenna, and Fons 2004). This impression may be achieved by presenting a narrative that claims to support a generally acceptable position, but that actually supports a more extreme and possibly objectionable position.

Neojihadism: The Jihadi Narrative

[President George W.] Bush left no room for doubts or media opinion. He stated clearly that this war is a Crusader war. He said this in front of the whole world so as to emphasize this fact. . . . When Bush says that, they try to cover up for him, then he said he didn't mean it. He said, "crusade." Bush divided the world into two; "either with us or with terrorism." . . . The odd thing about this is that he has taken the words right out of our mouths.

—Osama bin Laden, November 3, 2001

The role of narratives, as well as the approaches used to analyze them, is relatively novel in the field of terrorism studies. Nevertheless, studies on the topic have been fruitful in helping us understand how narratives can contribute to the persistence of terrorist ideology (Al Raffie 2012). Narratives play multiple critical roles in terrorist propaganda, seduction, and recruitment. The concepts of emotional appeals, personal identification, latitude of acceptance, identification, absorption, and transportation are all relevant to understanding terrorist narratives. Al-Qaeda's call to arms, for example, is a globally resonant set of narratives expressing its vision, grievances, agenda, and hopes, and has a proven ability to turn passive observers into active participants in violent extremism (Fink and Barclay 2013).

Recent research on radicalization (HSPI/CIAG 2007; Neumann 2012; Roy 2008a; Roy 2008b; Bergin, Osman, Ungerer, and Yasin 2009; Levitt 2009; Presidential Task Force 2009; Stevens and Neumann 2009) has focused increasingly on the narratives being developed by al-Qaeda and related groups. Thus, for example, the terrorist-promoted narrative of "oppression and victimhood" suffered by Muslims around the world at the hands of

Western Crusaders is regarded as the main fuel for the growth of extremism (Office for Security and Counterterrorism 2009, 141). Thus, understanding al-Qaeda's narrative is important because it constitutes a heuristic framework that offers guidance both to local extremist groups in relation to the objectives to be achieved and to individuals who are potential recruits or are on the path of self-radicalization, and also provides explanations for terrorist activities. An important contribution in this respect is that of Jeffrey Halverson, Steven Corman, and H. L. Goodall in their book *Master Narratives of Islamist Extremism* (2011), which posits that even though there are differences in locally embedded narratives, master narratives exist that override these local narratives in importance. There is no single "manifesto" or official source for the master narrative, although many would cite works such as Sayyid Qutb's *Milestones* (1964), Ayman al-Zawahiri's *Knights Under the Prophet's Banner* (2001), and Abdullah Azzam's *Defence of the Muslim Lands: The First Obligation after Faith* (1979).

It is possible to identify the "core" violent radical narrative, which includes the following elements:

1. During the time of the Prophet, there was a "Golden Age of Islam." This age, which was characterized by harmony, peace, and justice, forms the millennial vision to which violent radicals desire to "return."
2. Cultural pollution (accumulated cultural practices or innovation), moral decay, and Muslim "sinfulness" led to the end of the Golden Age.
3. Since that time, Muslims have suffered repression, ethnic cleansing, and "defeats" including the loss of statehood (the Ottoman Caliphate) in the early twentieth century; Palestinian displacement and the loss of Jerusalem in the mid-twentieth century; repression in areas such as North Africa, Egypt, and Chechnya; ethnic cleansing in the Balkans; infidel occupation of holy places in Saudi Arabia since the 1991 Gulf war; and the occupation of Afghanistan in 2001 and Iraq in 2003.
4. The United States, Israel, and the United Kingdom/European Union are seen as being united in a "Zionist"/"Crusader" axis, with Islam as the main enemy. The populations of these countries are seen as "collectively guilty" and therefore are not subject to the norm of civilian protection. This narrative has moved from engagement with the "near" enemy (e.g., a secular government in Egypt) to a globalized jihad against the "far enemy" (i.e., expelling the West from the Middle East).
5. The experience of defeats, seen as relevant to all Muslims as part of the global *ummah* (community), is combined with the perception of

a "Westernization" of Muslim life. The main elements of perceived poisonous Western influence include the corrupting ascendancy of the material over the spiritual life, permissive attitudes and behaviors, complacency, and comfort being valued more than honor or glory.

6. Islam is under general unjust attack by Western crusaders led by the United States. The jihadists, regarded as "terrorists" by the West, are defending against this attack, and the actions that they take in defense of Islam are proportionally just and religiously sanctified. Therefore, it is the duty of good Muslims to support these actions. (Betz 2008; Change Institute 2008, 45)

As this core narrative indicates, the master narrative combines four separate narratives (Leuprecht et al. 2010). The *political narrative* is concerned with the evils of the West, including a neo-Marxist take on global inequities arising from Western hegemony and exploitation. The *moral narrative* focuses on the internal contradictions of liberal democracies, which profess freedom as their core value and equality and justice as their subsidiary values, although these values are unrealizable ideals and indeed drivers of a society's moral decay. The *religious narrative* legitimizes violent struggle to defend Islam against the crusader West. Finally, the *social-psychological narrative* employs a classic in-group/out-group strategy to brand as infidels those who do not buy into this syllogism, while promoting the brotherhood of arms as a means of countering social exclusion and of fulfilling a yearning for adventure and sacrifice.

Indeed, as Christopher Heffelfinger (2010, 4) says, the central narrative put forward by al-Qaeda and its affiliates is that the Muslim world is under siege; Muslims are suffering violent onslaught, hardship, and oppression; and the only appropriate response is violent jihad. This narrative has been repeated by jihadist leaders and theologians in ideological tracts and in online publications promoted by al-Qaeda and its affiliates. The repetition, from a variety of sources, is essential to reinforcing the narrative's message. It is also backed by facts and examples: the Muslim casualties of the wars in Iraq and Afghanistan, the actions of repressive regimes in Muslim countries (which purportedly remain in power because of Western support), exploitation of Muslim wealth and natural resources, discrimination and bias against Muslim minorities, and other real or perceived grievances.

When Thomas Friedman tried to understand why Major Nidal Malik Hasan killed 13 innocent people at Fort Hood in November 2009, he came up with one answer: "[T]he more it seems that Major Hasan was just another angry jihadist spurred to action by 'The Narrative.' What is scary is that even though he was born, raised and educated in America, The Narrative still got to him" (Friedman 2009). According to Friedman (2009), "The Narrative is the cocktail of half-truths, propaganda and outright lies about America that have taken hold in the Arab-Muslim world since 9/11. Propagated by jihadist Web sites, mosque preachers, Arab intellectuals, satellite news stations and books—and tacitly endorsed by some Arab regimes—this narrative posits that America has declared war on Islam, as part of a grand 'American-Crusader-Zionist conspiracy' to keep Muslims down." A Jordanian expert, quoted by Friedman, argues, "This narrative is now omnipresent in Arab and Muslim communities in the region and in migrant communities around the world. These communities are bombarded with this narrative in huge doses and on a daily basis. [It says] the West, and right now mostly the U.S. and Israel, is single-handedly and completely responsible for all the grievances of the Arab and the Muslim worlds."

In her study of Salafi-jihadi narratives, Dina Al Raffie (2012) found similar elements: In the Salafi jihadism context, the worldwide suffering, endless humiliation, poverty, and oppression of many Muslims are not only the fault of the corrupt governments of these countries, but are also due to their deviation from Islam, which has resulted primarily from the brutal colonization of many of these countries. This is one example of many where the jihadists seek to group the grievances and frustrations of Muslims around the world and provide easy-to-grasp explanations for their misfortune; the appeal of their narratives a measure of the "simplicity of message and linkages with real-world grievances" (Lia 2008, 3). In his book *The Future of Political Islam*, Graham Fuller (2004, 1) adequately summarizes how the concept of humiliation is one of the key supporting elements of the master narrative:

> The deepest underlying source of Muslim anguish and frustration today lies in the dramatic decline of the Muslim World . . . from the leading civilization in the world for over one thousand years into a lagging, impotent and marginalized region of the world. This stunning reversal of fortune obsessively shapes the impulses underlying much contemporary Islamist rhetoric.

The so-called master narrative of Salafi jihadists has increasingly shifted its focus to not only foreign governments but also foreign populations, where the latter are pools for recruitment. The master narrative is aimed at radicalizing not only Muslims at "home" but also those from Muslim Diasporas in

North America and Europe. According to Al Raffie (2012, 17): "The master narrative focuses on the sufferings and grievances of fellow Muslims in one of two theaters: 1. Countries where Muslims are the minority and, 2. Muslim-majority countries that are perceived to be suffering heavy losses due to foreign intervention. The grievances of Muslims in both cases are caused by non-Muslim regimes and are thus framed as a war against Islam."

The Palestinian issue is also a cornerstone in many jihadi narratives since it relates to the existing hatred of Israel and, by extension, of Jews. Capitalizing on these negative emotions, which have become inherent in many Arab communities, al-Qaeda narratives advance an anti-Semitic worldview that frames current conflicts in a religious context in order to spread the rhetoric of a Jewish conspiracy aimed at the destruction of Islam and Muslims. According to a study of the content of jihadist websites following Operation Iraqi Freedom in 2003, the focal point of the narrative shifted to place increased emphasis on the notion of regional domination by the "Jewish-Crusader" alliance thereafter (Hegghamer 2006). According to Philipp Holtmann (2013), another aspect of al-Qaeda's master narrative is the "far enemy–near enemy" frame, according to which Islam is under attack at the same time by "corrupted Muslim governments and their collaborators" (the near enemy) and "the Zionized Neo-Crusader alliance" (the far enemy).

In terms of the seductive elements in narratives (e.g., identification, transformation), the master narrative has a strategic outlook, in that it works to create both real and perceived bonding or personal identification by promoting hostilities between Muslims and non-Muslims, and transforming the individual into the perception of a "War on Islam" (Al Raffie 2012, 19). These narrative elements are the most frequently mentioned appeals in our analysis of terrorist "chatter" and online postings, as well as by Al Raffie (2012, 19): "Offering itself up as a perfected, purified version of Islam, the Salafi Jihadist master narrative is the reflection of the broadest interpretations of Salafism in current geopolitics and acts as a platform for radicalization with the ultimate goal being the mobilization of action."

However, the al-Qaeda narratives are changing and vary according to location, purpose, target population, and time. The newer versions enrich the central narrative, neither rejecting nor replacing it. It is clear that the central narrative in their own view is based on the "clash of civilizations" between the West and Islam. Alex Schmid (2010, 46) describes al-Qaeda's single narrative as a "unifying framework of explanation that provides vulnerable Muslims with an emotionally satisfying story to make sense of the world in which they live and their role in it." The Internet with its numerous online platforms has been an ideal stage for the promotion of the central narrative. Magnus Ranstorp (2007, 31), for example, states that:

> In this virtual battlefield it is clear the militants have mastery of mechanisms to project this "single narrative" in a way that carries enduring resonance and with a logic that thousands of Muslims find absolutely compelling.

The new methods and tactics used to sell the jihadi master narrative should be related to the term "neojihadism," suggested recently by Peter Lentini (2013). Neojihadism, according to Lentini, is a multifaceted religious and political phenomenon that is based on a victimization narrative (which depicts Muslims as an oppressed and stigmatized community), advocates violence as the only means to end the victimization, and uses a set of interpretations of Islamic texts to justify this violence. It is both a subculture (which celebrates and shares violent actions through digital media and global narratives), and a counterculture (which is antagonistic toward mainstream Islam and toward the cultures and politics of Muslim-majority states and other countries with Muslim diaspora communities). Neojihadism, therefore, is situated at the end of a religious and ideological continuum that departs from earlier radical Islamists such as Sayyid Qutb, Muhammad Abd-al-Salam Faraj, and Abdullah Azzam, all of whom advocated violence through insurrection against agents of the state but did not advocate terrorism. Lentini's concept of neojihadism is wider than al-Qaeda's single narrative. Rather than being linked to a specific organization, neojihadism transcends al-Qaeda, and identifies the organization as part of a broader global movement that draws on this narrative to justify its use of violence and encourages others to do so.

To function, this neojihadist narrative relies mainly on the use of online platforms. The Internet is hardly the only instrument of recruitment and radicalization; yet its participatory culture and participatory architecture (which enables users to easily connect, produce, and distribute content online) create new opportunities to propagate terrorist narratives for groups, cells, and lone wolves. Moreover, the Internet's virtual communities of like-minded people can arise and potentially connect people across great distances, creating groups in which social dynamics may accelerate the spread and acceptance of the radical narrative.

Reception of the Terrorist Narratives

Clearly, the terrorist narrative never appears as "terrorist." It is often presented as a story or a set of consistent and coherent stories on history, on current events, on the future. How does this neojihadist narrative "sell" to Muslims? There is ample survey evidence to show that many Muslims in the United Kingdom and United States, as well as in Muslim countries, see the Global War on Terror as a war on Islam.[1] When asked, "In the aftermath of the September 11 attacks, do you feel the U.S. is fighting a war on terrorism or a war against Islam?," the percentage of American Muslims who affirmed "Islam" has kept on rising steadily from 18 percent in 2001, 31 percent in 2002, 38 percent in 2004, to 55 percent in 2007. Comparable results come also from the United Kingdom (Leuprecht et al. 2009). In a February 2009 poll of Muslim countries by the Centre for the Study of Terrorism and Responses to Terrorism (START), the proportion of respondents who thought that the United States has goals that are hostile to Islam ranged from 62 percent in Indonesia to 87 percent in Egypt (World Public Opinion 2009, 11). As Leuprecht and his colleagues (2009, 3) conclude: "Such findings confirm that the West is losing the narrative in the misnamed 'War on Terror'. Even more critically, however, some opinion polls indicate that the current strategy to develop a Western counter-narrative has not just failed but may actually be counter-productive." Consistent with the narrative that the war on terrorism is a war on Islam are Muslim opinions about the presence of US troops in Muslim countries. A 2009 START poll asked, "Overall, do you think the US having naval forces based in the Persian Gulf is a good idea or a bad idea," and those who answered "bad idea" ranged from 76 percent in Jordan to 91 percent in Egypt. A similar consensus emerges when respondents are asked whether they endorse the goal of al-Qaeda to "push the US to remove its bases and its military forces from all Islamic countries": 87 percent of Egyptian, 64 percent of Indonesian, and 60 percent of Pakistani respondents concur with this goal (World Public Opinion 2009, 7). Of course, not all of those who see Islam under attack would necessarily endorse terrorism as a legitimate response that warrants support. Indeed, considerable evidence from public opinion surveys indicates that

---

[1] See, for instance, Zogby International polling in the American Muslim Poll in conjunction with Georgetown University's Project on Muslims in the American Public Square (Project MAPS 2002), the Hamilton College Muslim America Poll (Hamilton College 2002), and Pew Research Center polling from "Muslim Americans: Middle Class and Mostly Mainstream" (Pew Research Center 2007).

most Muslims do not agree with terrorist tactics. Yet the "funnel" stage of the terrorist recruitment and radicalization (see chapter 3) requires that only a fraction of the targeted population would be receptive of the more extreme views. For example, in a July 2005 ICM Research telephone poll of British Muslims, only 5 percent of respondents answered "justified" to the question, "Do you think any further attacks by British suicide bombers in the UK are justified or unjustified?" Assuming that there are about one million adult Muslims in the United Kingdom, 5 percent works out to 50,000 Muslims (Leuprecht et al. 2009).

David Gartenstein-Ross and Laura Grossman, the authors of *Homegrown Terrorists in the U.S. and U.K.* (2009), studied radicalization of 117 homegrown "jihadist" terrorists from the United States and United Kingdom. From their study, the authors identified six characteristics that best express the radicalization of terrorists known to have participated in an attack or an attempted attack. To a varying degree, all of the homegrown terrorists possessed a legalistic interpretation of Islam; placed their trust in selected Islamic authorities; believed in an inherent schism between Islam and the West; had a low tolerance for theological deviance; attempted to impose their beliefs on other Muslims; and experienced political radicalization based on the beliefs that the West seeks to subjugate Islam, that most Muslims fall short of the true faith, and that military action represents the only proper Muslim response. The evidence of the adaption of the master narrative is clear:

> As homegrown terrorists radicalize, they often come to perceive an inherent schism between Islam and the West—believing that the two are at odds, and perhaps even incapable of coexistence. This perception can be expressed in a number of ways. In some cases, individuals attempt to isolate themselves from Western society physically. In others, these individuals will explain the perceived schism between Islam and the West to friends, family, or conspirators. (Gartenstein-Ross and Grossman 2009, 13)

The 2008 Change Institute report *Studies into Violent Radicalisation: The Beliefs, Ideologies and Narratives* reveals additional evidence regarding the terrorist narrative. The overall aim of this study was to explore those beliefs, narratives, and ideologies that lead to violent radicalism with a view to understanding of the causes and remedies for violent radicalization. The research was conducted through analysis and empirical data collection through fieldwork interviews in Denmark, France, Germany and the

United Kingdom, sampled to capture the diversity of the Muslim "field" in each country.

The Change Institute's research identified and analyzed the narratives of violent radicals and those who show sympathy or support for their "cause," and explored the ways in which people embrace these ideas and make them their own. The key narratives identified were living in a "hostile" society, disenfranchisement and heightened political consciousness, anti-imperialism and social justice, revivalism, emancipation, and the personal search to be a good Muslim. The study found that "the proto-ideology of violent radicals possesses explanatory power derived from the employment of legitimate 'mainstream' narratives and benefiting from a favorable context for the propagation of their ideas" (Change Institute 2008, 5). Moreover, the violent radicals were adapting central arguments of the master jihadi or Salafi narratives. As the report concludes:

> The analysis . . . serves to highlight how a range of ideologies and narratives have been adopted by individuals, groups, organizations and movements. Whilst the narratives identified are often of specific relevance to the context of the group in question, it is nevertheless possible to identify common characteristics in the development, application and role of the beliefs, ideologies and narratives in violent radical groups. The primary theme that is shared by all of the groups reviewed is that of a sense of grievance, perceived or real, supplemented by narratives of deliberate victimhood that informs the employment and justification of violence. These grievances, however misconstrued, are not just 'irrational' sentiments i.e. not subject to rational analysis, but are supported by ideological frameworks that outline the nature and cause of perceived problems, a vision of the future and a prescription for action. (Change Institute 2008, 32)

Frequently, the evidence of the success of terrorist narratives comes from the terrorists themselves. Reading the declarations of suicide bombers, the evidence from failed or caught terrorists, and the letters and documents left or sent by terrorists reinforces the impression that the deadly narratives find their way to some followers.

Adam Yahiye Gadahn

Perhaps the best illustrative example of the terrorist narrative's effectiveness is the case of Adam Yahiye Gadahn, an American who became a senior al-Qaeda operative, cultural interpreter, official spokesman, and media advisor. Gadahn was born in California in 1978 to a Jewish father and a Christian mother. In 1995, at age 16, Gadahn moved in with his grandparents in Santa Ana, California. While living with his grandparents, Gadahn described himself as having a "yawning emptiness" and sought ways "to fill that void." In 1995, at age 17, Gadahn began studying Islam and converted to Islam later that year. Gadahn reportedly moved to Pakistan in 1998, and in a short time became a senior advisor to Osama bin Laden, playing the role of "translator, video producer, and cultural interpreter" (Khatchadourian 2007). Gadahn himself describes the role of his exposure to online radical Islamist narratives: when he was 17 and still living with his grandparents, he posted an essay on the website of the University of Southern California's Muslim Students Association describing his conversion. In the essay, entitled "Becoming a Muslim," he described his first online revelations:

> The turning point, perhaps, was when I moved in with my grandparents here in Santa Ana, the county seat of Orange, California. My grandmother, a computer whiz, is hooked up to America Online and I have been scooting the information superhighway since January. But when I moved in, with the intent of finding a job (easier said than done), I begin to visit the religion folders on AOL and the Usenet newsgroups, where I found discussions on Islam to be the most intriguing. You see, I discovered that the beliefs and practices of this religion fit my personal theology and intellect as well as basic human logic. Islam presents God not as an anthropomorphic being but as an entity beyond human comprehension, transcendent of man, independent and undivided. (Gadahn 1995)

Starting in 2004, Gadahn appeared as Azzam al-Amriki ("Azzam the American") in a number of videos produced by al-Qaeda. A US court indicted Gadahn in absentia for treason in 2006. The Federal Bureau of Investigation (FBI) considers him one of the "most wanted terrorists" in the world, and the US State Department is offering a $1 million reward for his capture.

As mentioned above, in their study on the radicalization of homegrown American and British terrorists, Gartenstein-Ross and Grossman (2009) identified six manifestations of this process. These stages fit the Gadahn case well. Getting deeper into the radical Islamist milieu, he isolated himself from non-Muslim family members and effectively tried to block out the Western world. He lived in a drab apartment in which the only decorations were Islamic sayings of the Prophet Muhammad and a timetable for *salat* (prayers). As he adopted an increasingly legalistic view of Islam, some of his instructors guided him to a more radical path, lecturing him on the evils of the United States and Western society. He came to see Islam and the West as irreconcilably opposed. Eventually, Gadahn made his way to Pakistan, where Khalid Sheikh Mohammed, one of the masterminds of the 9/11 attacks, is believed to have recruited him. Gadahn's numerous videos and taped declarations promote the killing of Americans, jihad, and martyrdom, based on the neojihadist narrative that he now endorses: "Muslims in the West have to remember that they are perfectly placed to play an important and decisive part in the jihad against the Zionists and crusaders, and to do major damage to the enemies of Islam, waging war on their religion, sacred places and things and brethren. This is a golden opportunity and a blessing."[2]

Nidal Hasan

Major Nidal Hasan was 39 years old when he went on a shooting spree at the US Army base in Fort Hood, Texas, on November 5, 2009, killing 13 and wounding 29. Hasan had been radicalized for a number of years: he had no public associations with violent extremists, but made extensive use of the Internet. Between December 2008 and June 2009, he sent 20 emails to Anwar al-Awlaki, sharing his thoughts, expressing his admiration, and, importantly, seeking permission to carry out the attack. In one of these emails, Hasan wrote al-Awlaki: "I can't wait to join you" in the afterlife. Hasan also asked al-Awlaki when jihad is appropriate, and whether it is permissible if innocents are killed in a suicide attack. Accordingly, al-Awlaki, who responded to two of Hasan's emails, was described as a "virtual spiritual sanctioner" in the US Senate Homeland Security Committee's report on the shooting (Lieberman and Collins 2011, 20). In the months before the shooting, Hasan increased his contacts with al-Awlaki. Al-Awlaki had set up a website, with a blog on which he shared his views. On December 11, 2008, al-Awlaki had condemned any Muslim who seeks a religious decree "that would allow him to serve in the armies of the disbelievers and fight against his brothers." However, on November 9, 2009, four days after the Fort Hood shooting, al-Awlaki praised Hasan's actions in another post on his website:

---

[2] Adam Gadahn, in a video message released on the Internet, June 3, 2011.

Nidal Hassan is a hero. He is a man of conscience who could not bear living the contradiction of being a Muslim and serving in an army that is fighting against his own people. . . . Any decent Muslim cannot live, understanding properly his duties towards his Creator and his fellow Muslims, and yet serve as a US soldier. The U.S. is leading the war against terrorism which in reality is a war against Islam.

In March 2010, al-Qaeda spokesman Adam Yahiye Gadahn also praised Hasan. In a video message, he said that although Hasan was not a member of al-Qaeda:

[The] Mujahid brother . . . has shown us what one righteous Muslim with an assault rifle can do for his religion and brothers in faith. . . . [He] is a pioneer, a trailblazer and a role-model . . . and yearns to discharge his duty to Allah and play a part in the defense of Islam and Muslims against the savage, heartless and bloody Zionist Crusader assault on our religion, sacred places and homelands.

Arid Uka

In March 2011, Arid Uka killed two US airmen at Germany's Frankfurt airport (see chapter 7). During Uka's trial, it became clear that he had been radicalized within just a few weeks. Again, a seductive narrative played a significant role: The trigger for Uka's action was a YouTube video that showed Muslim women being raped by American soldiers. Uka thought that the film was real, but it turned out to be a scene from the 2007 Brian De Palma antiwar film *Redacted*, taken out of context. "I thought what I saw in that video these people would do in Afghanistan," Uka told the German court in August 2011 (BBC 2011). Uka was a lone perpetrator whose entire radicalization happened online. He was a member of several extremist forums and maintained a Facebook page under the name "Abu Reyyan," where he displayed links to the websites of many extremist websites and preachers. Uka himself described becoming increasingly introverted in the months before the attack, staying at home and playing computer games and watching Islamist propaganda on the Internet.

Colleen LaRose

Colleen LaRose, better known as "Jihad Jane," is an American citizen who was convicted and sentenced to 10 years for terrorism-related crimes, including conspiracy to commit murder in a foreign country and to provide material support to terrorists (Department of Justice 2010). LaRose converted to Islam in 2005, after she had begun reading Muslim websites and signed up at a Muslim dating site. She learned the basics of Islam from a mentor in Turkey and converted "via instant messenger." "I was finally where I belonged," she recalled. Her involvement in online jihad began in June 2008, when she started commenting on YouTube videos about conflicts in the Middle East. Based in a small town in Pennsylvania, she became immersed in jihadist websites and extremist forums. On June 20, 2008, she posted a comment on YouTube using the screen name "JihadJane," saying that she was "desperate to do something somehow to help" suffering Muslims. On MySpace and other social networking sites, she announced her desire to help Muslims and, beginning in December 2009, declared that she wanted to become a martyr, "dying in Allah's cause." One of her co-conspirators allegedly identified Lars Vilks, a Swedish artist who in 2007 had outraged some Muslims by drawing a cartoon of the Prophet Muhammad, as a target. LaRose was directed on March 22, 2009, to go to Sweden, find, and kill Vilks to frighten "the whole Kufar [non-believer] world" (Department of Justice 2010). According to her indictment, she responded in writing: "I will make this my goal till I achieve it or die trying" (see Shiffman 2010).

Terry Loewen

Terry Loewen, an avionics technician and Muslim convert, was arrested in December 2013 for attempting to bomb the Wichita Mid-Continent Airport in Kansas (Anti-Defamation League 2013). The FBI had been investigating Loewen for months, during which time he repeatedly expressed his desire to "please Allah" and to "be active in some kind of . . . jihad to feel I'm doing something proactive for the Ummah [community] . . ." The FBI agents, who posed as Al-Qaeda in the Arabian Peninsula (AQAP) operatives in the sting, communicated online with Loewen. Loewen was ready and fully committed to a "martyrdom operation" (suicide bombing) in which he and one of the "brothers" (FBI agents posing as al-Qaeda operatives) were to die in a truck laden with explosives on the tarmac of the Wichita airport when the maximum number of "target" aircraft were present with passengers. Loewen described in his own words the motive for his "mission":

As time goes on I care less and less about what other people think of me, or my views of Islam. I have been studying subjects like jihad, martyrdom operations, and Sharia law. I don't understand how you can read the Qur'an and the sunnah of the Prophet and not understand that jihad and the implementation of Sharia is absolutely demanded of all the Muslim Umma. (Quoted in Boyle and Associated Press 2013)

Loewen continued in a later communication to express his satisfaction with his decision to become a "shaheed" (martyr) for the cause of Allah and related it to the messages of the jihadi master narrative:

Inshallah, this operation will be huge. Just to be a part of any operation with these brothers is a great honor for me, but of (sic) it can instill and great financial loss to [those] who run this country, then I will truly feel blessed. . . . May this mission, inshallah, be fruitful for all of us. . . . By the time you read this I will—if everything went as planned—have been martyred in the path of Allah. There will have been an event at the airport which I am responsible for. The operation was timed to cause maximum carnage + death. . . . My only explanation is that I believe in jihad for the sake of Allah + for the sake of my Muslim brothers + sisters. (Gillam 2013)

The "Toronto 18"

The Canadian "Toronto 18" terror ring, apprehended in a series of arrests in the greater Toronto area in 2006, provides an example of a larger group of radicalized youth. Saad Gaya, a member in this group, was arrested in Toronto in June 2006. The 18-year-old was inside a warehouse, unloading a truck full of what he thought was ammonium nitrate fertilizer. Gaya had become convinced that detonating bombs in his home country was a rational response to Canada's military mission in Afghanistan: "I was young and politically naïve," Gaya told the judge, "Today, however, I recognize how irrational and unreasonable this line of thought was." A central member of the "Toronto 18," Gaya remains behind bars, serving an 18-year prison term. Saad Khalid, who was arrested with Gaya as they unloaded the fertilizer, turned to radical Islam after he developed a fierce "connection toward the Muslim cause" and an "increased sense of meaning." The group's ringleader, 20-year-old Zakaria Amara, dreamed of becoming an Islamic scholar. Amara's psychiatrist wrote: "Internet jihad videos became more exciting and their causes became more urgent." Fahim Ahmad, who also pleaded guilty to his role in the "Toronto 18" conspiracy, told the court that he spiraled into a "fantasy world" fueled by online instigators and a driving desire to feel "larger than life" (Friscolanti 2013).

The current and emerging narratives identified in this chapter are critical to understanding the full range and complexity of the new terrorism's online outreach. The roots of the "core" or master narrative can be traced back to a combination and cross-fertilization of beliefs and values from various sources, including Salafi theology and the ideology of the Egyptian Muslim Brotherhood and similar groups. The conflict against the Soviet Union in Afghanistan in the late 1980s provided the setting for these ideologies to evolve into the al-Qaeda narrative espoused by Osama bin Laden and Ayman al-Zawahiri, which has been promoted to an expanding Muslim diaspora. However, the situation becomes more complex, for in addition to the continuing contribution from al-Qaeda, both affiliated groups and unaffiliated organizations have been promoting similar or related narratives (Change Institute 2008, 33). An added complexity to the myriad sources is the "bottom-up" spread of narratives, supported by various Internet platforms.

Counternarratives

The reasons why an individual becomes radicalized are not yet fully understood. The fact that an individual embraces the terrorists' narrative does not necessarily mean that he or she will join the jihad against the West or actually engage in any terrorist activity. Yet progress in counterterrorism appears to be related to both establishing a credible narrative and damaging the terrorist narrative (Archetti 2012, 129). William Casebeer and James Russell (2005, 9), for example, suggest that the most effective way to counter terrorism is by developing a "better story" to replace "their" narrative. In the United Kingdom, a special cross-departmental communication unit, the Research, Information and Communication Unit, was set up in Whitehall in 2007. Its task was specifically to "use messaging to disrupt the Al Qa'ida narrative" (Office for Security and Counterterrorism 2009, 153). In the United States, a Presidential Task Force report (2009) also argues for "rewriting the narrative."

The construction of a counternarrative to violent jihad should be seen as part of a long-term strategy to combat radicalization and recruitment into Islamist militancy. The counternarrative differs from a counter-information campaign in that, "more than simply maligning the enemy or challenging its message, it offers an alternative vision to which one opts in; a storyline that gives meaning to the actions it is requesting of the subscriber." This narrative, must discredit that of the jihadists—most importantly by delegitimizing the violence they promote—while making a compelling case for forms of nonviolent activism and civic participation (Heffelfinger 2010, 3).

This is far from being an easy mission. The starting point is the state of the jihadi or neojihadist narratives versus the Western (or US) narratives of democracy, modernity, human rights, and the like, among target populations. The comparative evidence is not encouraging. One public opinion survey conducted in several Islamic countries (Kull et al. 2009) investigated popular attitudes toward al-Qaeda, terrorism, and US policies. Large majorities appear to know the al-Qaeda narrative, particularly its goals and its perspective on US foreign policy. Echoing the al-Qaeda worldview, majorities in several Muslim countries state that al-Qaeda's goals are establishing sharia law (76 percent in Pakistan; 76 percent in Morocco; 65 percent in Egypt; 49 percent in Indonesia); getting the United States to withdraw from Muslim countries (87 percent in Egypt; 72 percent in Morocco; 65 percent in Indonesia; 60 percent in Pakistan); and keeping Western values out of Muslim countries (88 percent in Egypt; 76 percent in Indonesia; 64 percent in Morocco; 60 percent in Pakistan). Likewise, America's goals are also perceived according to the al-Qaeda narrative framework: spreading Christianity (Pakistan: 71 percent; Morocco: 67 percent; Indonesia: 55 percent); weakening and dividing Islam (Egypt: 87 percent; Pakistan: 74 percent; Indonesia: 62 percent); maintaining control over oil (Egypt: 88 percent; Morocco: 82 percent; Indonesia: 67 percent; Pakistan: 62 percent). By contrast, only small minorities (13 percent in Indonesia; 10 percent in Pakistan; 8 percent in Egypt) think that the United States favors democracy in Muslim countries—a poor showing for the narrative actually promoted by the American administration. Moreover, public opinion does not believe that the United States respects international law (20 percent in Egypt; 20 percent in Pakistan; 12 percent in Indonesia), and instead believes that the United States is a hypocritical power that does not respect these laws (78 percent in Pakistan; 67 percent in Egypt; 55 percent in Indonesia) (see Archetti 2012, 159).

The speed and power of the online platforms and social media in spreading narratives and shaping perceptions make them a vital tool for fighting back. Former US secretary of defense Donald Rumsfeld has stated, "Today we are fighting the first war in the era of e-mail, blogs, BlackBerries, instant messaging, digital cameras, the Internet. . . . The US government still functions as a five-and-dime store in an eBay world" (Rumsfeld 2006). The good news is that the technologies and platforms that are being used to spread the ideology of terrorism can also be used to combat it. However, effective counternarrative strategy will require understanding of "strategic rhetoric."

Narratives, like any other stories, rely on rhetoric elements and especially on the three components offered by the ancient Greek philosopher Aristotle in his *Rhetoric*: (1) ethos (2) logos, and (3) pathos.

- *Ethos:* appeals that the communicator makes to the audience to establish credibility, to ensure that the audience can trust him or her.
- *Logos:* appeals to facts such as events, traumas, conflicts, deprivation, or inequalities.
- *Pathos:* appeals that the communicator makes to the audience's emotions, including anger, hate, pride, and frustration (quoted in Casebeer and Russell 2005, 11).

These components can be very useful for structuring counternarrative campaign. Based on the notion of *noise* in communication processes, all three elements may be used to both impair the terrorist narrative and promote an alternative narrative. The *ethos* component requires credible communication, fronted by communicators who have the personality and reputation required to ensure that the message will be received and believed. At the same time, on the offensive front, it may require "noise" operations that attack the credibility of the terrorist sources. Consideration of *logos* involves the rational elements of the narrative: Is it logical? Is it consistent enough to be believed? This logical reasoning may involve the need to find the right arguments, valid and accurate, based on irrefutable facts and supported by reliable sources. At the same time, the offensive tactic must attack the accuracy of facts and data delivered by the terrorist narrative. Finally, *pathos* requires the activation of emotional appeals, while countering those on which the terrorist narrative is based (Casebeer and Russell 2005).

There is an additional source for counterterrorist narratives: learning from those who have decided to *leave* terrorist organizations. As suggested by Michael Jacobson (2010a), in order to determine the kind of counternarrative that might be effective among those seemingly hardened individuals already in terrorist organizations or those well along the path to radicalization, it is useful to study people who have voluntarily walked away from these paths. Determining the reasons for such a change in perspective could help in crafting messages designed to pull people away from terrorist organizations. In the United Kingdom, former members of the radical group Hizb ut-Tahrir established the Quilliam Foundation, which describes itself as "Britain's first Muslim counter-extremism think tank" (quoted in Jacobson 2010a, 74). Quilliam aims to undermine the ideological foundation of radical extremism by refuting its premises.

As with the radicalization process, which seems to differ from person to person, there is not one dominant reason why individuals have walked away from terrorism. The reasons can be strikingly prosaic: family, money, petty grievances. But they can also revolve around shaken ideology or lost faith in a group's leadership. Counterterrorism agencies may be able to take advantage of the knowledge of these trends to better formulate appropriate counternarratives. Led by lessons from cases of terrorist dropouts, Jacobson (2010a, 75–78) suggests several counternarrative motives:

1. *Undermine terrorist leadership:* From the various terrorist dropout cases, it seems clear that a general lack of respect for a group's leadership has often been a factor in dropping out of a terrorist group or path. Thus, undermining terrorist and extremist leadership should constitute one part of the tactic. Crafting messages that significantly detract from leaders' authority and credibility is vital.

2. *Highlight civilian/Muslim suffering, hypocrisy of the Islamist narrative:* An effective counternarrative should also demonstrate civilian and Muslim victimization by extremism and terrorism. Disillusionment with terrorists' strategy and actions has been found to play a major part in why people have left such groups.[3]

3. *Portray terrorists as criminals:* Many terrorist groups, including al-Qaeda, are increasingly involved in a variety of criminal activities. These include a wide array of criminal activity, ranging from cigarette smuggling to selling counterfeit products, from identity thefts to production and selling of drugs. According to the Drug Enforcement Administration, 19 of the 43 US-designated foreign terrorist organizations are definitively linked to the global drug trade, and up to 60 percent of terrorist organizations are suspected of having some ties with the illegal narcotics trade (Braun 2008). Painting terrorists as common criminals may help demonstrate the impurity of their motives, ideology, and supposed religious conviction.

---

[3] One example is Omar bin Laden, Osama bin Laden's fourth son. He had spent nearly five years living in Afghan training camps but, following 9/11, Omar quit al-Qaeda and called its attacks "craziness," according to journalist Peter Bergen. He continued, "[t]hose guys are dummies. They have destroyed everything, and for nothing. What did we get from September 11?" (quoted in Bergen 2007).

4. *Focus on life as a terrorist:* The reality of life for a terrorist has often driven people out of their organizations. Through studies of the personal stories of terrorist dropouts, it can be discerned that the individual operatives' perceived lack of respect from leaders was influential in their decision to break from the radical group. If people are joining because the terrorist lifestyle seems glamorous or because they believe they are fulfilling some larger purpose, demonstrating the reality will help to dispel these myths. This may involve the use of former members who can describe their unsatisfying lives as members of a terrorist organization, emphasizing that it simply does not live up to the hype.

A crucial aspect in this counternarrative campaign is the presenter of the message. Governments are not the most effective messengers for presenting the counternarrative. Other actors may make more effective and credible presenters. These actors may include former terrorists, family members of terrorists, and religious authorities. A narrative presented by former terrorists would resonate strongly with their audiences, especially because they can deliver particularly strong messages about the reality of life as a terrorist and effectively leverage their disillusionment with the cause to lure both potential and active terrorists away from extremist groups. Family members of terrorists can also play an important role in trying to persuade individuals to leave terrorist organizations; after renewed contact with their families, many trained terrorists subsequently decided to abandon the plots in which they had been selected to participate (Jacobson 2010b, 78–79). Finally, religious leaders and respected scholars may be effective counterterrorism messengers. This is a particularly important part of the engagement process in Muslim communities, and it is vital that the target community should consider the leader or scholar to be a credible and impartial source of information. In September 2013, the Qatar International Academy for Security Studies released the results of a year-long research project that sought to identify concepts and strategies for dealing with violent extremism. One of the findings was that religious leaders and community resiliency groups can play an important role in both countering extremist narratives and rehabilitating extremists (The Soufan Group 2013).

The answer to the question of whose voice is most effective in terms of delivering a counternarrative depends on which audience one wants to reach. Each group, region, country, and community requires a unique approach to countering the call to terrorism.

Is it possible? The art of persuasion is well known and widely used in the West. It can be harnessed to the countercampaigning efforts. For instance, Western societies are well acquainted with commercial and political campaigns. These campaigns have produced a wealth of accumulated know-how and experience, as well as professionals and experts in the domains of sophisticated advertising and public relations. Moreover, the complexity of the challenge should involve the use of scientific methods and knowledge, from social psychology to experimental designs. Empirical findings from the numerous studies on persuasive communication may be highly relevant. Even more relevant are the sophisticated methods used to test and improve the persuasive messages. A vast range of experimental designs can be applied to test the impact of persuasive contents, appeals, arguments, visuals, and more. They are activated on experimental audiences, composed of groups based on social characteristics that resemble those of the target populations. Countercampaigns will require new counterterrorism "armies" that have new strategies, capabilities, tactics, and cyberweapons to counteract the extremist and violent narratives and persuasive tactics (Sinai 2006). Intense interagency, intergovernmental, and international communication and harmonizing processes, embedded in an institutional framework and with clear defined rules of the game, are required to make such a campaign effective and efficient.

The "Say No to Terror" Campaign

The "Say No to Terror" (SNTT) campaign may serve as an illustrative example of a counter campaign relying on counternarratives (Aly, Weimann-Saks, and Weimann 2014). The SNTT campaign is a comprehensive online social marketing campaign, consisting of a website and social media platforms,[4] which uses a variety of mechanisms to communicate a counternarrative to selected elements of the terrorist narrative. The website is entirely in Arabic and hosts information content (Mission Statement/About Us) as well as short videos, forums, posters, and links to social media platforms (Facebook, You-Tube, and Twitter). Users who register on the website can post comments

---

[4] See SNTT website (http://www.sntt.me/), Facebook (https://www.facebook.com/saynototerror), Twitter (https://twitter.com/saynototerror), YouTube (http://www.youtube.com/user/saynototerror), all retrieved on March 15, 2014.

about the videos as well as other material. The SNTT campaign is specifically aimed at a Muslim Arabic audience underscored by the campaign's
slogan "Terrorism. I am Muslim: I am against it." According to the website,
"Terrorism is a criminal act targeting innocent people, and it deserves to be
fought by all means and to have its claims and its devastating effects on our
society disclosed." The "About Us" section of the full SNTT website states:

> We believe in the justice of true Islam which calls for solidarity and
> mutual assistance, for the support of the oppressed, for tying the hands
> of the oppressor, and for spreading the good word in order to restore the
> shine of Islam which beamed his light to the world and whose teachings
> influenced civilizations in the East as well as in the West. We look at the
> Muslim world today and we realize that our Islam is threatened from
> within by those who are pretending to be defending religion while reli-
> gion has nothing to do with them.
>
> A group of agitators are attacking society; their aim is to deceive our
> brothers and sons and they maim the image and message of Islam and our
> tolerant sharia by their criminal actions. We defend the greatness of Islam
> and the purity of its sons and their sincere affiliation with peoples and
> tribes created by Allah in order to get to know one another, coexist and
> cooperate in the peopling the Earth and preserving the human dignity.
>
> Based on all of the above, our mission aims to expose the claims of
> terrorist agitators and unveil their crimes, to encourage all those who
> have a conscience to reject their criminal acts and destructive ideas, and
> to fight them in order to protect our society from their wrongs and their
> destructive impact on all levels.

Apart from what is stated above, little information is offered about the website creators, their origins, or their affiliation. A close analysis of the website
content suggests that it is sympathetic to Saudi Arabia. Posts that refer to specific religious issues (such as *takfir*—accusing other Muslims of apostasy)
or situations (such as the Syrian conflict) are consistent with Saudi Arabia's
stated position on matters pertaining to the Arab Spring and Western intervention in Arab affairs. These posts give precedence to the Saudi Arabian
monarchy as "The Custodian of the Two Holy Mosques" and the ultimate
authority in Islam, and expose the monarchy's efforts to counter terrorism
(Aly, Weimann-Saks, and Weimann 2014).

An analysis of 15 videos on the SNTT website revealed salient themes
and examined how these themes construct an effective counternarrative

(Aly, Weimann-Saks, and Weimann 2014). The analysis highlighted the following themes:

1. The detrimental consequences of joining a terrorist group for the individual and their family;
2. Terrorist groups use manipulation and lies to influence others;
3. Terrorists are the "enemies of Islam" who kill innocent Muslims (including children);
4. Muslims have a duty to be vigilant against terrorism and to protect themselves and their communities from extremism.

These themes are constructed around familiar concepts in traditional Muslim culture. The videos themselves do not overtly challenge ideological assumptions or religious interpretations of Islamic concepts such as jihad or *takfir*, although they do use Quranic verses and popular hadiths to accompany images. Rather, they rely on Muslim traditions that value family and the collective good over individualism. In this respect, the videos present a narrative that mirrors elements of the terroristic narrative. Terroristic narratives online commonly construct the call to armed jihad as an obligation for all Muslims: those who heed the call and fight for the "oppressed" are hailed as martyrs, and their altruism serves as an inspiration to the broader community of Muslims to also take up arms. For example, the video entitled "Awakening the Conscience" warns against the detrimental consequences of replacing one's family with the "terrorist family." A young man from a middle-class family somewhere in the Middle East is shown on his way back to his neighborhood, as locals seem to recognize him. He seems to be lost and his expression is disillusioned. He has memory flashbacks: a young boy being trained militarily by a sheikh, scenes of executions of Muslim people, money in the hands of leaders of jihadist groups, food feasts enjoyed by these leaders and from which the young man has been excluded; he then remembered how he used to sit with his family around the dining table and how his mother used to pay particular attention to him. The video concludes with the following text, a verse from the Quran: "Extremism is a road leading up to delusion. Do not throw yourselves with your own hands into destruction." Similarly, the video "The Plight of a Family" portrays a father disowning his son who has joined a terrorist organization, and the video "The Friend" calls on family members and friends to intervene if a family member is exhibiting signs of siding with terrorism, focusing heavily on the strength of family bonds within Arab culture.

Drawing on Muslim social constructs of collective good, several other videos urge viewers to be vigilant about terrorism and its influence. In the video "An Eye That Watches Is Better Than an Eye That Cries," a mother weeps as she watches her son on a jihadist video posted on the Internet. The film warns viewers: "The Internet is a way to communicate and a gateway to knowledge. But the terrorists also see it as a window/path to our children, to brainwash their young minds and to convince them of their criminal principles. Our duty is to protect our children from danger and deception, not only in the schools and on the roads, but also within the sanctity of our homes. Terrorists are determined to mislead our children." Another video, "The Clowns," also calls for vigilance but targets public support for terrorism by encouraging viewers to speak out against terrorism. Two films, "Zakat" and "Good Charity/Bad Charity," attempt to raise public awareness of terrorist financing operations that pose as valid charities. Both films call on viewers to take personal responsibility for ensuring that charitable donations and *zakat* (Muslim obligatory charity) do not end up funding terrorist activities.

Finally, several videos focus on the theme of terrorism as victimizing Muslims. "The Enemy Within" shows images from the scene of a terrorist attack in an undisclosed Muslim country. Each victim of the attack is named in an attempt to draw attention to the humanity of the victims, and perhaps also to their Muslim religion. The text accompanying the images reads: "Thousands of innocent people die as victims of misguided terrorists pretending to act in the name of Islam." The videos "The Scream," "I Am Innocent of Your Crimes," and "No Life Flourishes Where There Is Terrorism" all draw on images of innocence juxtaposed against images of terrorism. These three videos use visuals of children and/or related images (e.g., toys, teddy bears) to draw attention to the detrimental impact of terrorism. Although some parts of the videos refer explicitly to the killing of innocent Muslim children by acts of terrorism, others use childlike imagery as a symbol of the future to highlight the futility of terrorism. In "The Scream," a young toddler is held hostage by a group of terrorists. The imagery is accompanied by the following voice-over: "Tomorrow is made up of every child today. So do we raise the child in a safe and healthy society, according to the teachings of Islamic tolerance, justice, freedom, equality and exchanging advice? Or do we leave him hostage in the hands of extremism, injustice, *takfir*, kidnapping, and destruction?"

There is no empirical evidence regarding the success of the SNTT campaign (Aly, Weimann-Saks, and Weimann 2014). The indicators of high exposure are impressive—several SNTT videos on YouTube have enjoyed more than 700,000 views—but views alone are not valid indicators of effects or impact. Thus, although the SNTT campaign demonstrates an attempt to launch a counterterrorist narratives campaign, using the same platforms that terrorists use and targeting the same audiences, the impact of such efforts is still to be studied and measured.

Chapter 11

Challenging Civil Liberties

Those who would give up essential liberty to purchase a little temporary safety deserve neither liberty nor safety.

—Benjamin Franklin, 1759

Since September 11, 2001, many governments have sought to address concerns that terrorists are using the Internet and online platforms for communicative and instrumental purposes. These concerns have led to several countermeasures that have caused fears and worries among civil liberties activists. For example, the organization Reporters Without Borders (Reporters Sans Frontières) argues,

Several Western democracies have become "predators of digital freedoms," using the fight against terrorism to increase surveillance on the Internet. . . . A year after the tragic events in New York and Washington, the Internet can be included on the list of "collateral damage." . . .

Cyber liberty has been undermined and fundamental digital freedoms have been amputated." (Quoted in Associated Press 2002)

As terrorists' use of the new media became more intensive, sophisticated, and alarming, so did the various countermeasures launched by governments, military, and counterterrorist agencies. The war on terrorism allowed for the application of intrusion, eavesdropping . . . but why should law-abiding citizens be affected by counterterrorism measures on the Internet? Computer databases store the most intimate details of our daily lives, including medical records, banking and investment transactions, credit reports, employment records, credit card purchases, photographs, and fingerprints. Surveillance cameras are ubiquitous at banks, airports, and other public places. Over the years, courts and lawmakers have sought to protect this private information from unnecessary disclosure and to enact strict guidelines on the interception and gathering of such information by government agencies. How can we balance the need for Internet security with minimal cost in terms of civil liberties?

Post 9/11 Counterterrorism Measures on the Net

Less than a week after the 9/11 attacks, several legislative steps were introduced to minimize the risk of additional terrorist attacks. Forty-five days later, President Bush signed the USA PATRIOT Act, a legislative step that increased the surveillance and monitoring capabilities of law enforcement agencies. The Electronic Privacy Information Center (2005) criticizes the act, saying, "Though the Act makes significant amendments to over 15 important statutes, it was introduced with great haste and passed with little debate, and without a House, Senate, or conference report. As a result, it lacks background and legislative history that often retrospectively provides necessary statutory interpretation." The act was in fact a legislative step intended to strengthen the nation's defense against terrorism, including in its provisions the problematic monitoring of private communications and access to personal information. The USA PATRIOT Act included a so-called sunset provision, according to which certain "sections of the act automatically expire after a certain period of time, unless they are explicitly renewed by Congress" (Electronic Privacy Information Center 2005). Some of the sunset provisions concern electronic surveillance and the Federal Bureau of Investigation's (FBI) use of an Internet-monitoring system called Carnivore.

The act of capturing Internet traffic is known as "sniffing"; the "sniffer" is the software that searches the traffic and grabs items that it is programmed to find. Intrusion detection systems use sniffers to match transmitted data, including email messages, against a set of rules. Law enforcement agencies that need to monitor email during an investigation may use a sniffer designed to capture extremely specific traffic. Following the 9/11 attacks, the FBI unveiled the sniffer known as Carnivore, which was already in use. The FBI explains the origin of the code name: "Carnivore chews all the data on the network, but it only actually eats the information authorized by a court order" (Graham 2001). (Carnivore was later renamed DCS1000 to prevent it from sounding too much like a privacy-consuming predator.) According to the FBI, Carnivore is much like the common Internet monitoring tools and commercial "sniffers" used daily by many Internet companies, Internet service providers (ISPs), and monitoring agencies. It operates like a telephone wiretap applied to the Internet, examining each of the exchanged packets and recording those that relate to suspicious issues. Since most Internet traffic in the United States flows through large ISPs, the agencies install "sniffers" like the Carnivore inside the ISPs' data centers. It is very likely that Carnivore or DCS1000 have been replaced by newer technologies and methods, as revealed in 2013 by former National Security Agency (NSA) operator Edward Snowden. Snowden's leaked documents uncovered the existence of numerous global surveillance programs, many of them run by the NSA with the cooperation of telecommunication companies and European governments. They revealed the existence of the Boundless Informant data mining tool, the PRISM electronic data mining program, the XKeyscore analytical tool, the Tempora interception project, the MUSCULAR access point, and the FASCIA database, which contains trillions of device-location records. In their book *Counterstrike: The Untold Story of America's Secret Campaign Against Al Qaeda*, Eric Schmitt and Thom Shanker (2011) describe counterattacks on terrorist websites and online platforms.[1] The book, in chapters called "Terror 2.0" and "The New Network Warfare," sheds light on offensive US cyber operations almost never discussed by US officials. According to the book, the US military hacked and temporarily disabled Iraqi insurgent and terrorist-based websites. At least two sites were "knocked off the web" prior to Iraq's March 2010 national election. The sites, including one sponsored by a "shadowy organization" called the

---

[1] See also Joshua Sinai's article "Terrorism on the Internet and Effective Countermeasures" (2011).

JRTN,[2] "were posting specific operational information that was considered a clear and emerging threat to the security of the vote," the authors wrote. At least one site, hosted by a US-based Internet service provider, was shut down after a visit from US lawyers "presenting snapshots of virulent, extremist and violent web pages carried on their server." (Schmitt and Shanker 2011, 139–40). The US provider was not identified.

Monitoring the Internet raises the issue of overcoming encrypted messages. Encryption is software that locks computerized information to keep it private; only those with an "electronic key" can decode the information. Finding a way to crack encryption has typically baffled law enforcement agencies. According to legal expert Matthew Parker Voors (2003, 348), "[i]f the government discovered a suspicious e-mail that was encrypted, and wanted to read it, it had two options—it could obtain the private-key from the sender, or it could attempt to break the code through a brute force attack. The first option, requiring terrorists to supply the private-key, is not plausible because this would reveal the investigation to the terrorists. . . . The second option, cracking the code by a brute force attack, is possible, but the process involves a massive amount of computer power and an equally large number of staff hours." Neither option is attractive, so a third option evolved: in 2001, the FBI confirmed the introduction of its key logger system, code-named "Magic Lantern" (Sullivan 2001). Magic Lantern is a program that, once installed on a suspect's computer, records every keystroke typed. These gathered keystrokes are then analyzed by the FBI to find passwords, and using these "harvested" passwords, the FBI can access the suspect's email messages and documents and even the computers contacted by the suspect. Thus, Magic Lantern allows the FBI to record a suspect's keystrokes and learn his private encryption key. And yet the FBI is not "stealing" the key, but rather records it only after a process of authorization and when the individual involved is a suspect in terrorist activity. Therefore, FBI-developed tools like the Magic Lantern, some argue, are designed to protect civil liberties more than is usually done in commercial surveillance tools.

In 2003, a revised version of the USA PATRIOT Act was prepared. This version, the Domestic Security Enhancement Act of 2003 (informally labeled "PATRIOT II"), expanded surveillance power, increased government access to private data, and broadened the definition of terrorist activities. The major criticism of PATRIOT II came from the American Civil

---

[2] JRTN stands for *Jaysh Rijal al-Tariqa al-Naqshbandia* (Army of the Men of the Naqshbandi Order, or Naqshbandi Army), a Sufi insurgent group in Iraq.

Liberties Union (ACLU). The ACLU argued, "The new 'anti-terrorism' legislation goes further than the USA PATRIOT Act in eroding checks and balances on presidential power and contains a number of measures that are of questionable effectiveness, but are sure to infringe on civil liberties" (ACLU 2003). However, proponents of the act argued that monitoring the Internet by eavesdropping on email and phone calls was intended to help prevent crime and fight terrorism. Terrorists and criminals can be tracked better than ever before because their online activities, such as using a credit card, sending an email message, booking a flight, or paying a toll, leave an electronic trail (Schwartz 2001).

In April 2004, an email message intercepted by NSA investigators led to a massive investigation conducted by intelligence officials of several countries. This investigation ended with the arrest of nine men in the United Kingdom and one in Ontario, Canada, who were later charged with facilitating a terrorist act and being part of a terrorist group (Akin 2004). This was the first time that the American regular monitoring of email traffic led to an arrest. Behind the monitoring and the arrest was the NSA system of surveillance. A year later, a broader scope of the NSA's activities was exposed, starting the lingering debate on its legitimacy.

The NSA Controversy

All wiretapping of American citizens by the NSA requires a warrant from a three-judge court set up under the Foreign Intelligence Surveillance Act (FISA).[3] After the 9/11 attacks, Congress passed the USA PATRIOT Act, which granted the President broad powers to fight a war against terrorism. The George W. Bush administration used these powers to bypass the FISA court and directed the NSA to spy directly on al-Qaeda in a new NSA electronic surveillance program. The full scope of the program was secret, but through it the NSA had total, unsupervised access to all fiber-optic communications between some of the nation's largest telecommunication companies' major interconnected locations, including phone conversations, email, web browsing, and corporate private network traffic. News reports

---

[3] The Foreign Intelligence Surveillance Act of 1978 (FISA) is a US federal law that prescribes procedures for the physical and electronic surveillance and collection of "foreign intelligence information" between "foreign powers" and "agents of foreign powers" (which may include American citizens and permanent residents suspected of espionage or terrorism). The law does not apply outside the United States. It has been repeatedly amended since the 9/11 attacks.

in December 2005 first revealed that the NSA had been intercepting Americans' phone calls and Internet communications. When the NSA's spying program was first exposed by the *New York Times* in 2005 (Risen and Lichtblau 2005), President Bush admitted to part of the program, which allowed the NSA to monitor, without the use of warrants, the communications of between 500 and 1,000 people inside the United States who had suspected connections to al-Qaeda. But other facets of the program were aimed not just at targeted individuals, but at perhaps millions of innocent Americans who had never been suspected of a crime. Elected officials, civil rights activists, legal scholars, and the general public all found the program's legality and constitutionality—and the potential for abuse—to be cause for concern. The controversy has since expanded to include the media's role in exposing a classified national security program.

How was the NSA able to monitor the communication traffic? First, the government convinced major US telecommunications companies such as AT&T and Sprint to hand over their customers' "call-detail records," which included names, street addresses, and other personal information. In addition, the government received detailed records of the calls that customers had made to family members, coworkers, business contacts, and others. Second, the same telecommunications companies also allowed the NSA to install sophisticated communications surveillance equipment at key telecommunications facilities around the country (Electronic Frontier Foundation 2014). This equipment gave the NSA unfettered access to large streams of domestic and international communications in real time, in what amounted to at least 1.7 billion emails a day, according to the *Washington Post* (Priest and Arkin 2010). The NSA could then mine the data and analyze this traffic for suspicious keywords, patterns, and connections. In March 2012, James Bamford reported that the NSA was spending $2 billion to construct a data center in a remote part of Utah to house the information it had been collecting for the past decade. "Flowing through its servers and routers and stored in near-bottomless databases," Bamford (2012) wrote, "will be all forms of communication, including the complete contents of private emails, cell phone calls, and Google searches, as well as all sorts of personal data trails—parking receipts, travel itineraries, bookstore purchases, and other digital 'pocket litter.'"

Moreover, as revealed by former NSA contractor Edward Snowden, US and British security agencies found ways to unlock encryption used to protect emails, banking, and medical records. The agencies, Snowden's documents revealed, adopted numerous methods that include covert measures to ensure NSA control over the setting of international encryption standards, the use of supercomputers to break encryption, and—the most closely guarded secret of all—collaboration with technology companies and ISPs themselves. Through these covert partnerships, the agencies have inserted secret vulnerabilities, known as backdoors or trapdoors, into commercial encryption software. Snowden also revealed that the NSA spends $250 million a year on a program which, among other goals, works with technology companies to "covertly influence" their product designs. GCHQ (Government Communications Headquarters), the British equivalent of the NSA, had been working to develop ways into encrypted traffic on the "big four" service providers, named as Hotmail, Google, Yahoo!, and Facebook. The agencies insist that their ability to defeat encryption is crucial to their mission of counterterrorism (Ball, Borger, and Greenwald 2013).

The NSA controversy has several dimensions: one of the major issues is that of the warrantless surveillance or "warrantless wiretapping" (Sanger and O'Neil 2006). Under this program, referred to by the Bush administration as the Terrorist Surveillance Program (part of the broader President's Surveillance Program), the NSA was authorized by executive order to monitor, without search warrants, the phone calls, Internet activity (e.g., web, email), text messaging, and other communication involving any party that the NSA believed to be outside the United States, even if the other end of the communication was within the United States. However, it has been discovered that all US communications have been digitally cloned by government agencies. The excuse given to avoid litigation was that no data hoarded would be reviewed until searching it would be defensible. Under public pressure, the Bush administration allegedly ceased the warrantless wiretapping program in January 2007 and returned review of surveillance to FISA. Thus, in 2008 Congress passed the FISA Amendments Act of 2008, which relaxed some of the original FISA court requirements. During the Obama administration, the NSA has allegedly continued operating under the new FISA guidelines. However, in April 2009, officials at the US Department of Justice acknowledged that the NSA had engaged in an "overcollection" of domestic communications in excess of the FISA court's authority (Risen and Lichtblau 2009).

In summer 2013, public controversy arose in response to Snowden's unauthorized revelations of the agency's unprecedented capacity to spy on the private communications of US citizens. News reports in the international media have revealed questionable operational details about the NSA. The reports emanated from a cache of top-secret documents leaked by Snowden.

In addition to US federal documents, Snowden's cache reportedly contained thousands of Australian, British, and Canadian intelligence files. In June 2013, the first of Snowden's documents were published simultaneously by the *Washington Post* and the *Guardian*, attracting considerable international attention. A significant portion of the full cache of the estimated 1.7 million documents was later obtained and published by media outlets worldwide (Strohm and Wilber 2014). Facing growing criticism, President Obama made a public appearance on national television on August 6, 2013, where he reassured Americans that "we don't have a domestic spying program" and "there is no spying on Americans." The extent to which the media reports have responsibly informed the public is also disputed. In January 2014, Obama said that "the sensational way in which these disclosures have come out has often shed more heat than light," and tried to explain the role of the NSA's eavesdropping:

> Americans recognized that we had to adapt to a world in which a bomb could be built in a basement and our electric grid could be shut down by operators an ocean away. . . . So we demanded that our intelligence community improve its capabilities and that law enforcement change practices to focus more on preventing attacks before they happen rather than prosecuting terrorists after an attack. . . . Today, new capabilities allow intelligence agencies to track who a terrorist is in contact with and follow the trail of his travel or his funding. New laws allow information to be collected and shared more quickly and effectively between federal agencies and state and local law enforcement. Relationships with foreign intelligence services have expanded and our capacity to repel cyber attacks have been strengthened. And taken together, these efforts have prevented multiple attacks and saved innocent lives—not just here in the United States, but around the globe. (Obama 2014)

However, new revelations about the NSA's activities raised new concerns. In December 2013, the German magazine *Der Spiegel* revealed that the NSA operates a division known as Tailored Access Operations (TAO), an elite team of hackers that specialize in stealing data from the toughest of targets. This unit, according to the report, has intercepted computer deliveries, exploited hardware vulnerabilities, and even hijacked Microsoft's internal reporting system to spy on its targets. Citing internal NSA documents, the magazine said that TAO had a catalog of high-tech gadgets for particularly hard-to-crack cases, including computer monitor cables specially modified to record what is being typed across the screen, USB sticks secretly fitted with radio transmitters to broadcast stolen data over the airwaves, and fake base stations intended to intercept mobile phone signals on the go (Satter 2013). In January 2014, the *Guardian* reported that the NSA has collected and stored almost 200 million text messages per day from across the globe, using them to extract data such as location, contact networks, and credit card details. The NSA program, codenamed Dishfire, collected millions of text messages daily in an "untargeted" global sweep. According to the *Guardian*, "The NSA has made extensive use of its vast text message database to extract information on people's travel plans, contact books, financial transactions and more—including individuals under no suspicion of illegal activity" (Ball 2014). On January 17, 2014, President Obama announced that he would be appointing John Podesta "to lead a comprehensive review of big data and privacy" in the aftermath of revelations about the astonishing scope of the NSA's electronic spying programs (quoted in McCarthy 2014).

Balancing Security with Liberty

Fighting terrorism raises the issue of countermeasures and their cost. As terrorism experts John Arquilla and David Ronfeldt (2001, 14) argued in their article "The Advent of Netwar," "Terrorist tactics focus attention on the importance of information and communications for the functioning of democratic institutions; debates about how terrorist threats undermine democratic practices may revolve around freedom of information issues." Responding to the challenge presented by terrorism on the Internet is an extremely complicated and sensitive issue, since most of the rhetoric disseminated on the Internet is considered protected speech under the US Constitution's First Amendment and under similar provisions in other societies. The measures used and being considered could change the balance between privacy and security. In January 2014, President Obama spoke about the needed reforms to NSA programs:

> In our rush to respond to a very real and novel set of threats, the risk of government overreach, the possibility that we lose some of our core liberties in pursuit of security also became more pronounced. . . . The combination of increased digital information and powerful supercomputers

offers intelligence agencies the possibility of sifting through massive amounts of bulk data to identify patterns or pursue leads that may thwart impending threats. It's a powerful tool. But the government collection and storage of such bulk data also creates a potential for abuse. . . . So in the absence of institutional requirements for regular debate and oversight that is public as well as private or classified, the danger of government overreach becomes more acute. And this is particularly true when surveillance technology and our reliance on digital information is evolving much faster than our laws. (Obama 2014)

The need to protect the public from the threats of modern terrorism opens the door to the use of surveillance. Investigation and surveillance processes cause immense concern in contemporary society because they create and maintain social distinctions as certain groups are subjected to repressive measures for purposes of social control. David Lyon, sociologist and author of *Surveillance Society* (2001), acknowledges that while such practices are portrayed as protective of order and security, we should be asking whose purposes they serve. And how do we know that these ostensible purposes *are* served? Lyon emphasizes that surveillance is a tool that reinforces social and economic divisions, channels choices, and directs desires; it constrains, controls, and proves seriously problematic in issues of privacy.

Some people argue that, paradoxically, we must give up some freedoms in order to enjoy the ones we cherish the most. Amitai Etzioni, author of *The Limits of Privacy* (1999), argues in his 2002 essay "Seeking Middle Ground on Privacy vs. Security" that it is wrong to define counterterrorism surveillance tools as good or evil: "As with all technologies, the proper question is how it will be used. For instance, if evidence about a suspected terrorist is presented to a court of law, Magic Lantern should be allowed to decode the suspect's messages. But if it is installed at the discretion of every cop on the beat, the rights of many innocent people could be violated." Proper guidelines are key, and if these are not established, then such measures will usurp a large part of our liberties. Surveying the public, argues Etzioni, may well be unavoidable in the post–September 11 world, but he also warns that "retaining information about innocent conduct by innocent people poses a massive threat to privacy" (Etzioni 2002).

There are two concerns regarding the new digital war on terrorism. The first concern is that the new surveillance measures may improve security but harm civil liberties. The second concern relates to the very essence of what we mean today by privacy, and reflects a more general worry about the "retreat of privacy" resulting from the use of many high-tech surveillance tools. Although pre-9/11 legislation aimed at protecting people's privacy from invasion has been shelved, new antiterrorism laws give the authorities broad new powers to wiretap, monitor, or otherwise eavesdrop on Internet activity. The policies proposed and the practices applied since 9/11 are significantly accelerating the loss of civil liberties. These developments could have profound implications for democracies and their values, exacting a heavy price in civil liberties and increasing the destructive effects of terrorism in cyberspace. In the war on modern terrorism, timely information is more valuable than guns or missiles in saving lives. The question is how to gather that information efficiently without adopting the "draconian methods" of the terrorists' extreme ideologies (Lewis 2001, 200). President Obama, aware of this need to find the balance, declared in January 2014:

> I indicated in a speech at the National Defense University last May [2013] that we needed a more robust public discussion about the balance between security and liberty. Of course, what I did not know at the time is that within weeks of my speech an avalanche of unauthorized disclosures would spark controversies at home and abroad that have continued to this day. . . . We have to make some important decisions about how to protect ourselves and sustain our leadership in the world while upholding the civil liberties and privacy protections our ideals and our Constitution require. We need to do so not only because it is right but because the challenges posed by threats like terrorism and proliferation and cyberattacks are not going away any time soon. They are going to continue to be a major problem. And for our intelligence community to be effective over the long haul, we must maintain the trust of the America people and people around the world. (Obama 2014)

The "Golden Path"

All extremes are dangerous. It is best to keep in the middle of the road, in the common ruts, however muddy.

—Virginia Woolf, *The Common Reader*

If an elastic band is overstretched, noted Lindsey Wade (2003), eventually it will snap back with a sharp sting to the hand that forced it. Similarly, intensifying repression of civil liberties and exploitation of privacy are far more sinister in the long run than the threat of terrorism, international or domestic. We should recognize that terrorism has been around for hundreds of years and is not likely to go away. Modern societies will have to learn to live with some form of terrorism and with the threat of more sophisticated forms of terrorism, including cyberattacks and online radicalization, training, and launching of future terrorists. Countermeasures may be needed, but what are the civil-liberties versus terrorism-risk trade-offs? In 2003, W. Kip Viscusi and Richard Zeckhauser conducted a study to examine "people's willingness to sacrifice civil liberties in an effort to reduce terrorism risks, and also to explore aspects of individuals' terrorism risk perceptions that govern the character of their responses" (Viscusi and Zeckhauser 2003, 99). Viscusi and Zeckhauser argued that the desired balance between these conflicting concerns depends in large part on individual attitudes. The researchers surveyed a sample of Americans for their willingness to trade safety for civil rights. To measure the willingness trade-offs, they examined civil liberty issues pertaining to the targeting of passengers for screening at airports based on their demographic characteristics, most often salient characteristics such as ethnic background and country of origin; and surveillance of private mail, email, and phone communications. In particular, the respondents considered the following question: "Would you support policies that make it easier for legal authorities to read mail, email, or tap phones without a person's knowledge so long as it was related to preventing terrorism?" (Viscusi and Zeckhauser 2003, 107). According to the study's report, the findings revealed a willingness to sacrifice some civil liberties for increased security, which reflects the more general argument—articulated by Louis Kaplow and Steven Shavell (2002)—that many legal rights and liberties are not absolutes. As Viscusi and Zeckhauser (2003, 102) stated:

> Civil liberties and the prevention of terrorism are over two attributes for which society often makes extreme symbolic commitments toward the highest level. Many would argue that civil liberties are guaranteed rights, rights that cannot be compromised. In much the same way, advocates of risk control often claim that so long as any individual is at risk of being killed involuntarily, the risk must be reduced to ensure that we are in fact truly safe. Taken to the logical limit, this leads to the zero-risk mentality that pervades many legislative mandates of U.S. government risk and environmental regulation agencies, and is reflected in public risk attitudes as well.

In June 2013, a survey conducted by the Pew Research Center and the *Washington Post* found that a majority of Americans (56 percent) said that the NSA's program for tracking the telephone records of millions of Americans was an acceptable way for the government to investigate terrorism, though a substantial minority (41 percent) said that it was unacceptable (Pew Research Center for the People and the Press 2013). Although the public is more evenly divided over the government's monitoring of email and other online activities to prevent possible terrorism, these views are largely unchanged since 2002, shortly after the 9/11 terrorist attacks. The national survey conducted among 1,004 adults from June 6 to 9, 2003, found no indications that the latest revelations of the government's collection of phone records and Internet data have altered fundamental public views about the trade-off between investigating possible terrorism and protecting personal privacy. Sixty-two percent said that it is more important for the federal government to investigate possible terrorist threats, even if doing so intrudes on personal privacy, whereas only 34 percent said that it is more important for the government not to intrude on personal privacy, even if it limits the government's ability to investigate possible terrorist threats. These opinions have changed little since an ABC News/*Washington Post* survey in January 2006. These opinions have only small partisan differences: 69 percent of Democrats said that it is more important for the government to investigate terrorist threats, even at the expense of personal privacy, as did 62 percent of Republicans and 59 percent of independents. Even though there are apparent differences between the NSA surveillance programs under the Bush and Obama administrations, overall public reactions to both incidents are similar. In the 2013 survey, 56 percent said that it is acceptable that the NSA "has been getting secret court orders to track telephone calls of millions of Americans in an effort to investigate terrorism" (Pew Research Center for the People and the Press 2013).

President Obama's January 2014 speech, which outlined changes to the NSA's collection of telephone and Internet data, had little public impact: a January 2014 survey conducted by the Pew Research Center and *USA Today* found that overall approval of the program has declined since July 2013. In the January 2014 survey, 40 percent approved of the government's collection of telephone and Internet data as part of antiterrorism efforts, while 53 percent disapproved. In July 2013, more Americans approved (50 percent) than disapproved (44 percent) of the program. In addition, nearly half (48 percent) of respondents in the July 2013 survey said that there are not adequate limits on what telephone and Internet data the government can collect; in the January 2014 survey, even fewer (41 percent) said that there are adequate limits on the government's data collection (Klein and Soltas 2014).

These survey findings show that in the public's view, the optimal level of civil liberties, in practice, is not necessarily always the fullest extent of those liberties: Americans are willing to trade a degree of civil liberty for other valued benefits, such as the prevention of terrorism. The optimal level of civil liberties varies according to circumstances. For example, the public would not normally tolerate having every car driving along a roadway stopped and inspected, but would be more understanding if a serial killer were on the loose. Thus, a more realistic way to protect the Internet, to prevent its abuse by terrorists while at the same time protecting civil liberties, is to look for the "golden path," that is, the best compromise. The twelfth-century Jewish philosopher Mosheh ben Maimon (Moses Maimonides) termed the "golden path" the Right Path, a balance between two diametrically opposed extremes. Finding such a path means that we will have to accept both some vulnerabilities of the Internet to terrorism and some constraints on civil liberties, but the underlying guidelines should be to minimize both sorts of ills by finding the trade-offs between securing our safety and securing our liberties. Certain principles seem to take into account this balance between security and liberty within today's cyber-reality:

- Modifications of procedures and legislative acts
- Self-policing
- International collaboration.

It is worth investigating these principles in greater detail.

Modifications of Procedures and Legislative Acts

We will reform programs and procedures in place to provide greater transparency to our surveillance activities and fortify the safeguards that protect the privacy of US persons.

—President Barack Obama, January 2014

In the short time that the Internet has been publicly available (since the early 1990s) and the even shorter time that it has been widely used (since the late 1990s), it has revolutionized, for better or for worse, multiple aspects of lives in all corners of the world: commerce, communications, education, entertainment, and politics. Couple the Internet's ubiquitous and extensive influence with the unprecedented rate of development in Internet technology, and there is little doubt that the medium will continue to evolve and affect the world's populations. Never before have policymakers or the general public had to deal with such an evolution in their midst. Any laws, policies, or technologies intended to influence the development of the Internet must have visions for the possibilities and appreciate the probabilities for an Internet universe of the future. The flexibility and adjustability of laws pale in comparison to the flexibility of Internet technologies to adapt to new scenarios. Therefore, laws and policies addressing Internet uses and abuses must be crafted carefully and judiciously.

The USA PATRIOT Act and other legislative acts were designed and approved in the aftermath of the shocking events of 9/11. Now, more than a decade later, we can reexamine these measures, learn lessons from our efforts, and attempt to refine the laws and their implementation. One needed modification is to transform these laws into a "public pact" between society and the administration and security agencies. We are willing to submit ourselves to the security procedures of the Transportation Security Administration (TSA) in American airports because the procedures are based on agreed-upon trade-offs (our privacy and time in return for a reduced risk of being victimized while flying). Moreover, we know who the officials allowed to perform the search are, we know the search routines, and we know the limits of the procedures. Similar public understanding and acceptance should be applied to online counterterrorism measures. We will let the authorized officials search us and order us to take off shoes, coats, or belts, but we do it based on an agreed-upon "pact" that generally excludes further measures. We know that the American public supports the monitoring of the Internet, including private email traffic, but as in the case of TSA measures this acceptance relies on known procedures, known agents, and agreed-upon limits. If "sniffer" technologies are to be used, we need to know who is using them, how they are regulated, what guarantees are in place to limit their use to antiterrorism operations only, what kind of information is stored, and who has access to such information. As with airline travel, a pact is needed to regulate the authorities' access to private online communications. The option of curtailing NSA's bulk surveillance program altogether and instead using other means to gather information on suspected terrorists appears to be unrealistic. Thus, as in the TSA's example, clear regulations are needed on the following issues:

1. Who and what will be monitored online.
2. Who will be authorized to monitor and archive the information.
3. What data or information is to be archived.
4. Who will have the access to the archived information.

If these are the issues, what are the principles that would guide the decisions on their regulation? There is certainly a need to form such regulatory principles. Such reform has been suggested by several Internet companies, including AOL, Google, Facebook, Yahoo!, Twitter, Apple, LinkedIn, Dropbox, and Microsoft. In their November 2013 call to reform government surveillance, they stated:

> We understand that governments have a duty to protect their citizens. But this summer's [2013] revelations highlighted the urgent need to reform government surveillance practices worldwide. The balance in many countries has tipped too far in favor of the state and away from the rights of the individual—rights that are enshrined in our Constitution. This undermines the freedoms we all cherish. It's time for a change. . . . We urge the US to take the lead and make reforms that ensure that government surveillance efforts are clearly restricted by law, proportionate to the risks, transparent and subject to independent oversight. (Reform Government Surveillance 2013)

This call laid out the principles that should guide the regulation of online surveillance:

1. *Restraining governments' authority to collect users' information:* Governments should codify sensible limitations on their ability to compel service providers to disclose user data that balance their need for the data in limited circumstances with privacy interests and free flow of information. In addition, governments should limit surveillance to specific, known users for lawful purposes, and should not undertake bulk data collection of Internet communications.
2. *Accountability:* Intelligence agencies seeking to collect information should do so under a clear legal framework in which executive powers are subject to strong checks and balances. Reviewing courts should be independent and include an adversarial process.

3. *Transparency:* Transparency is essential to a constant examination of governments' surveillance powers and the scope of programs that are administered under those powers. Governments should allow companies to publish the government demands for user data and the rationale behind them.

4. *Free flow of information:* While trying to identify terrorist online traffic, its content and disseminators, governments should consider the need for free flow of information, permit the transfer of data, and should not inhibit access by companies or individuals to lawfully available information.

5. *Go global:* There are no borders in cyberspace; the Internet is a global arena. Any form of counterterrorism measures is restricted by neither national boundaries nor citizenship. Thus, it is not a matter of US policies or regulations: there is a need to reach an international collaboration, if not agreement, on standards and principles governing the use of online surveillance. (Reform Government Surveillance 2013)

The Obama administration has already begun to review NSA procedures in reaction to public outrage. The administration had asked US intelligence agencies and the Justice Department to come up with alternatives that would regulate data ownership and access. There are at least three viable options. The first option would be to keep the data in the hands of the Internet companies, including Facebook, Twitter, and YouTube. The NSA would then request access to specific records based on any connection to terrorists. However, the companies would certainly be against this proposal because the legal burden of turning over the data would still fall on them. The second option would be to warehouse the data with a government agency other than the NSA, such as the FBI or the Foreign Intelligence Surveillance Court. A third option would see the data turned over to a party other than the government or the Internet companies. But the last two options may end up as just an extension of the NSA. Moreover, all three options do little to forestall privacy fears; the bulk data would still be retained, but would merely change hands (Gorman and Barrett 2014). Each of the options has its own pitfalls, and none is likely to satisfy everyone. However, it is clear that data collection, storage, and analysis are crucial for the various agencies to prevent future terrorist attacks. If so, regardless of the option chosen, the need to regulate the process itself (who monitors, what is to be monitored, what is to be archived, and who has access to the archived data)—is also crucial.

Self-Policing

Most of the terrorist websites, forums, and chatrooms are posted on sites provided in Western, democratic, liberal countries. Almost all of the new social media platforms, from Facebook and Twitter to Instagram and You-Tube, are managed and owned by Western companies, mostly American. According to studies conducted by the Middle East Media Research Institute (MEMRI), 76 percent of the terrorist websites are hosted in the United States, with only 8 percent of sites hosted in the Middle East (Boccara 2004; Boccara and Greenberg 2004). The American-hosted sites include the websites of Hamas; Al-Aqsa Martyrs Brigades; the Palestinian Islamic Jihad and its military wing the Al-Quds Brigades; the Iraq Sunni insurgent group Army of Ansar Al-Sunnah; known supporters of al-Qaeda, including one that publishes al-Qaeda's *Al-Battar Training Camp* magazine; and a pro-Hezbollah weekly magazine.

In May 2013, the *Washington Times* reported that US Web companies are still hosting sites linked to terror. An Internet server in New Jersey, not far from 9/11's Ground Zero, hosted a jihadist leader's website that instructed supporters of al-Qaeda on how to use explosive devices against Western civilians, and provided blueprints that showed how to build the bombs. Another website hosted on a server in Miami provided Hezbollah—a State Department–designated terrorist organization—with a platform for its television website al-Manar (Carter 2013). In December 2012, MEMRI had noted that al-Manar was also using servers in the Netherlands and the United Kingdom. The Hezbollah site was immediately removed from the Dutch servers, but it is still using a server in the United Kingdom. Less than a month after being kicked out of the Netherlands, Hezbollah moved its television website to a US server (Stalinsky 2012b). Followers of Hezbollah could also view the terrorist organization's television programming on smart phones with the help of the Internet application WhatsApp, based in Dallas, Texas. YouTube and Twitter are also major platforms for Hezbollah's news outlets and videos. American hosting servers are a popular choice for practical and legal reasons. Domain names can be registered for as little as $11.99 per year, and hosting services with global bandwidths can be rented for a few dollars a month from companies like GoDaddy and Dynadot. The Dynadot server, located in San Mateo, California, offers a privacy service that allows registrants to mask their identity by listing addresses as "care of" the company—a convenience that has made it particularly popular with jihadists and Internet activists hoping to elude the authorities (Carter 2013).

In 2012, MEMRI used the WHOIS database to reveal how terrorist groups have come to depend on American companies for their activities online. When a new domain name is registered, the domain owner is required to submit information to the WHOIS database; however, some domain owners opt to use domain privacy services offered by registrars. With this option, they are technically compliant, but it enables them to avoid listing their actual information. In 2009, when radical Yemeni-American cleric Anwar al-Awlaki posted an article on his website praising Fort Hood shooter Nidal Hasan, MEMRI published an online report that included the WHOIS info for al-Awlaki's site. The report stated: "The registration information for his website is as follows: The domain ANWAR-ALAWLAKI.COM is protected by the private domain registration company DomainsByProxy.com." Within two hours of the MEMRI report's online publication, the site was removed and has not returned. The MEMRI report shows how US-based domain registration protection companies hide the identities of the individuals behind the most important al-Qaeda–affiliated websites and forums (Stalinsky 2012c). Namecheap, a company that provides a service called WhoisGuard to block domain registration information from showing up in database listings, is highlighted as hosting the two most important al-Qaeda–affiliated forums, Al-Shumoukh and Al-Fida. The MEMRI report named other US-based companies that provide protective services to the online activities of radical Islamic groups, including Network Solutions, Privacypost, Privacyprotect, Register.com, and PrivacyRegContact (Stalinsky 2012c).

Over the years, when a certain domain-name registrar is revealed to be hosting one of these sites, they are often promptly removed. Still, many continue to exist, and some speculate that the US government allows them to exist in order to monitor them. Usually, though, US-based ISPs host suspect websites without even knowing it. For example, an Internet company based in Burlington, Vermont, hosted a website that taught its members how to outfit a suicide bomber, aired al-Qaeda propaganda videos, and offered an "exclusive" Taliban video showing the beheadings of three "spies," according to computer records. The English-language website, Leemedia.net, was taken down—but not in a counterterrorism crackdown. Instead, the web server, Endurance International Corp. Inc., shut it down after Internet watchdogs made repeated demands to remove the terrorist material. The case of Leemedia.net, which was operated by a suspected terrorist sympathizer in Karachi, is an example of how US Internet companies may unknowingly host possibly hundreds of the most virulent Islamic extremist websites in the world, inciting young Muslims to kill Christians and Jews (Bender 2008).

The fact that all Islamist/jihadi and other terrorist websites are hosted directly or through subservers by Western—primarily American—ISPs raises the question of what can be done about it. In her *CQ Global Researcher* piece on terrorism and the Internet, Barbara Mantel (2009) noted that most ISPs, web hosts, and file-sharing and social networking sites have terms-of-service agreements that prohibit certain content. Yahoo!'s small business web hosting service, for example, forbids users from utilizing the service to provide material support or resources to an organization designated by the US government as a foreign terrorist organization. To that extent, there is an element of self-regulation within the information society. Because most of these websites are in Arabic, the language gap prevents most ISPs from knowing the content of the websites they host, but ISPs can certainly use the services offered by several organizations (including MEMRI) to provide ISPs with information about the content of the sites they host, so they can make an informed decision on whether they want to continue hosting these sites. ISPs retain the right to terminate service under any circumstances and without prior notice, especially if the content violates the terms of service agreement, or if law enforcement or other government agencies request its removal. Some ISPs reserve the right to remove information that is untrue; inaccurate; not current or incomplete; or is reasonably suspected of being untrue, inaccurate, not current, or incomplete. However, if the ISP does not remove such content, neither it nor its partners assume any liability.

Self-policing requires exposure: first, to expose the extremist sites and to inform ISPs and the public at large of their content, and second, to bring legal measures against ISPs that continue to host extremist websites and forums. Such implementations would require establishing a database—governmental or nongovernmental—that would regularly publish information about terrorist sites, forums, platforms, and social media, and provide it to ISPs upon request. Self-policing can be based on agreed-upon principles such as those suggested by the European Commission. In 2010, the European Commission called upon European Union member-states to submit project proposals to tackle the problem of online terrorism and to include public-private cooperation in the working method. This prompted the Dutch Ministry of Security and Justice to submit the Clean IT Project (2012) proposal. The project aimed to start a constructive dialogue among governments, businesses, and civil society to explore how to reduce the terrorist use of the Internet. The suggested principles, listed in the Clean IT Project's 2013 report, included the following:

Challenge 1: Terms of Use

Not all Internet companies state clearly in their terms and conditions that they will not tolerate terrorist use of the Internet on their platforms, nor do they define terrorism. This makes it more difficult to decide what to do when they are confronted with potential cases of terrorist incitement, recruitment, and training on their platform.

Practice: Internet companies can define and/or give examples of what is terrorist use of their services, and do so for legal, ethical, or business reasons. It is recommended that companies have sufficiently staffed and capable abuse departments and are consistent and transparent in how they deal with abuse of their networks and violations of their terms and conditions.

Challenge 2: Flagging

Internet users currently do not have enough easy ways of reporting terrorist use of social media. In addition, Internet users are not used to reporting what they believe is illegal. As a consequence, some terrorist use of the Internet is currently not brought to the attention of Internet companies and competent authorities.

Practice: Flagging is a useful method of notifying Internet companies about potential terrorist use of the Internet. User-friendly flagging systems have a separate, specific category to flag cases of terrorism. The service providers should also explain to their users how these flagging systems work and otherwise stimulate its use. This practice is primarily meant for social media or websites that provide user-generated content, but it could be considered to make the technology more widely available where this is technologically possible. Flagged content means that possible illegal content is brought to the attention of the service provider, and from that point they are not excluded from liability for the information stored on their networks.[4]

---

[4] As noted earlier, in 2013 an experimental study revealed that the effectiveness of the flagging system is rather limited: Out of 125 videos flagged, 57 (45.4 percent) were still online more than four months later (Stalinsky and Zweig 2013).

Challenge 3: End-User Mechanism

While content portals (like social networks, image, or video portals) can offer "flagging" opportunities, other platforms (like hosted websites) often lack such a mechanism. Moreover, there is not one international, user-friendly reporting mechanism available to all Internet users, irrespective of which part of the Internet they are using at the moment they notice what they think is terrorist use of the Internet.

Practice: A more systematic approach to help Internet companies to be notified by Internet users about alleged terrorist use of the Internet is a reporting mechanism that is implemented in the standard distribution of a browser, or, as a fallback solution only, is offered as a plugin for browsers. This is a user-friendly notification tool to Internet companies that do not offer flagging tools or do not have effective abuse departments. This mechanism should also be considered for the browsers of mobile devices and their operating systems.

Challenge 4: Referral Unit

Internet companies do have potential cases of terrorist use of the Internet reported to them, but they often lack the required specialist knowledge about terrorism to determine whether it is illegal. Determining what is illegal is primarily a law enforcement role. In other cases, Internet companies lack the language skills they need to make a judgment on the meaning and therefore on the legality of the content or other terrorist activity. As a consequence, a large number of potential cases of terrorist use of the Internet are not dealt with adequately.

Practice: Well-organized referral units and hotlines, having an appropriate team behind them with the needed competences and skills, will help ISPs to handle notifications about terrorist use of the Internet more effectively and efficiently. This is especially the case if ISPs are not sure whether possible terrorist use of the Internet that is reported to them is illegal or not.

Challenge 5: Point of Contact

Governments, Internet companies, competent authorities, and nongovernmental organizations do not always know who to contact on the issue of terrorist use of the Internet.

Practice: A network of trusted and listed points of contact facilitates cooperation between organizations committed to reducing the terrorist use of the Internet. Points of contact are experts able to represent their organization, preferably on a daily or even 24/7 basis. To establish a professional system of points of contacts, detailed working procedures and a central database will be required. These points of contact should be identified by role and their contact details published.

Challenge 6: Sharing Abuse Information

Most Internet companies have to deal with few cases of terrorism on their platforms. When illegal content is removed, terrorists often try and succeed to post it on other Internet companies' services.

Practice: Internet companies should share information on various forms of abuse of their network with each other, using a trusted intermediate partner organization. This private sector practice could be extended to include confirmed illegal terrorist use of the Internet. Only data that is formally confirmed as terrorist use of the Internet, taking into account national legislation, including privacy and data protection legislation, should be added to these sharing systems. (Clean IT Project 2013, 18, 20–22, 24–25)

International Collaboration

The war on terrorism is not a local war or even an American or Western war. It is a human, global, transnational war against unjustified use of violence, despicable and illegal conduct of conflict, and mass victimization of innocent civilians. To combat terrorism, there must be general international cooperation. This need applies also to the new arena of the Internet. Any attempt to limit terrorists' ability to access and use the Internet must be internationally based. The global nature of modern terrorism and the even more global nature of its use of the Internet require an international front to fight terrorists' abuse of the Internet. Without such collaboration, terrorists will simply move their sites from one country to another, seeking those hosts that will not interfere with their Internet activities. American efforts to fight Internet terrorism are carefully watched and even replicated or implemented by other societies. Yet if the struggle is limited to the United States, the war is lost. As we have seen, terrorist organizations find virtual shelter in various countries in Europe, the Far East, the Middle East, Russia, and Africa.

Defending the Internet on an international front is not merely a practical issue: The abuse of the Internet must be declared a multifaceted global threat. Moreover, this international front to defend the Internet must also protect the very nature of the medium and its future. This international front might take several forms, ranging from cooperation among nations to an international organization supported by many nations. Robert Stevenson (2003, 21) argues that the Internet adheres to the weakest-link principle: "If it's possible in any country, that becomes the standard around the world. In most cases, we applaud because it puts the authoritarian government on the defensive and makes censorship difficult. But sometimes, the weakest link becomes a global standard to the detriment of free expression as we define it" Stevenson (2003, 22) also describes several cases of transnational limits on Internet contents:

> In 1998, the head of the German subsidiary of the Internet provider CompuServe was convicted in a Bavarian court for failure to block child pornography and computer games with swastikas that had been distributed in user groups. The conviction was later overturned. Two years later, a French judge ordered Yahoo to remove Nazi paraphernalia from its auction website after a complaint was brought by groups devoted to fighting anti-Semitism and racism. The judge ordered Yahoo, which is based in the United States, to prohibit French access to the material. Inciting racism is a crime in France, and selling Nazi material, the judge said, offended "the collective memory of the country." Yahoo later got a ruling from a U.S. court that French law had no jurisdiction over its activities in the United States, including the posting of materials that are illegal in France. Still, Yahoo and other Internet-based companies became more sensitive to laws in other countries where they had an important presence and pulled materials that could be offensive—and potentially the basis of litigation—in other countries.

These few cases underscore the differences between how the United States and most other Western democracies deal with the Internet. For example, regarding the question of whether ISPs that host discussion groups and massive websites are responsible for those sites' content, the American answer is usually no, whereas in other countries the answer is typically yes. Says Stevenson (2003, 22–23), "In the United States, the answer is clearly in favor of free expression, even in issues involving child pornography or national security. In most other countries, it is usually the privacy and good name of individuals and the stability and order of the nation that take priority."

Given today's political environment, there is, of course, little or no chance for a global agreement, since political cleavages will prevent a wall-to-wall global front and leave several countries outside such a coalition. However, the more states that join this front and collaborate, the smaller the opportunities for terrorists will be. Today, most terrorist sites use Western servers, networks, and ISPs. Thus, Western societies should lead the international effort to fight terrorism on the Internet by seeking measures that will significantly limit terrorists' access to this medium and hamper their abuse of its liberal character.

In 2012, the United Nations Office on Drugs and Crime (UNODC) published the report *The Use of the Internet for Terrorist Purposes*. The report (UNODC 2012, v) states, "Despite increasing international recognition of the threat posed by terrorists' use of the Internet in recent years, there is currently no universal instrument specifically addressing this pervasive facet of terrorist activity. Moreover, there is limited specialized training available on the legal and practical aspects of the investigation and prosecution of terrorism cases involving the use of the Internet." It highlights the need to develop integrated, international, and specialized knowledge to respond to the continually evolving threat of online terrorism, and stresses international cooperation:

> Terrorist use of the Internet is a transnational problem, requiring an integrated response across borders and among national criminal justice systems. The United Nations plays a pivotal role in this regard, facilitating discussion and the sharing of good practices among Member States, as well as the building of consensus on common approaches to combating the use of the Internet for terrorist purposes. . . . There is currently no comprehensive United Nations treaty on terrorism that is applicable to an exhaustive list of the manifestations of terrorism. (UNODC 2012, 15, 18)

However, the international community has yet to agree on an internationally binding definition of the term "terrorism," owing largely to the difficulty of devising a universally acceptable legal categorization for acts of terrorism. If we can overcome this definitional problem, the UNDOC document describes certain international legal frameworks to fight online terrorism. For example, the case of terrorist incitement online: The crime of inciting terrorist acts is the subject of the 2005 United Nations Security Council resolution 1624. In that resolution, the Council called upon all states to adopt such measures as may be necessary and appropriate and in accordance with their obligations under international law to prohibit by law incitement to commit a terrorist act, and to prevent such conduct. However, cases involving statements by persons made over the Internet—especially when the alleged offender, the Internet services they use, and their intended audience are located in different jurisdictions—are regulated by different national laws and constitutional safeguards, and therefore present additional challenges for investigators and prosecutors from an international cooperation perspective (UNODC 2012, 135, 37).

As the UNDOC report reveals, international experience relating to the incitement to commit terrorist acts highlights two issues: first, how important (and sometimes difficult) it is in practice to differentiate between terrorist propaganda (statements advocating particular ideological, religious, or political views) from material or statements that amount to incitement to commit violent terrorist acts; and second, how the enforcement of laws dealing with alleged acts of incitement requires a careful case-by-case assessment of the circumstances and context to determine whether the institution of a prosecution for an incitement offence is appropriate in a particular case (UNODC 2012, 37). Yet, international agreement has been achieved, at least among European states. In Europe, article 3 of the Council of the European Union of November 28, 2008, on combating terrorism, and article 5 of the Council of Europe Convention on the Prevention of Terrorism oblige their respective member states to criminalize acts or statements constituting incitement to commit acts of terrorism. The Council of Europe Convention on the Prevention of Terrorism imposes an obligation on member states to criminalize "public provocation to commit a terrorist offence," as well as both terrorist recruitment and training. When calling upon states to criminalize the incitement of terrorist acts, UN Security Council resolution 1624 expressly provides that states must ensure that any measures adopted to implement their obligations comply with all their obligations under international law, particularly human rights, refugee, and humanitarian laws (UNODC 2012, 41, 44).

There is no doubt that the war on online terrorism relies on international cooperation. The universal instruments against terrorism in general, composed of international conventions and protocols and relevant resolutions of the UN Security Council, contain comprehensive mechanisms for international cooperation in criminal proceedings related to terrorism. These instruments make provisions for extradition, mutual legal assistance, transfer of criminal proceedings and convicted persons, reciprocal enforcement of judgments, freezing and seizure of assets, and exchange of information between law enforcement agencies. But none of these deals relates specifically to Internet-related acts of terror. It is clear that international cooperation in the investigation and prosecution of terrorism cases involving terrorist use of the Internet is hindered, to some extent, by the absence of a universal instrument dealing specifically with cyberterrorism. But cyberspace has no borders, no nations, and no boundaries. Thus, a terrorist attack may originate from one nation but target other nations. Cyberattacks against banks in Europe may have a catastrophic impact on the world economy, or a cyberattack on electricity plants in Canada may lead to a blackout in New York. In the absence of a counterterrorism instrument that deals specifically with Internet issues connected to terrorism, investigating and prosecuting authorities will continue to be reliant upon existing international or regional treaties or arrangements established to facilitate international cooperation in the investigation and prosecution of terrorism.

Public-Private Partnerships

". . . I want to just say this about the private sector. In my mind, the government is incapable of responding to its maximum ability without private sector support."

—Hon. Tom Ridge, former secretary,
US Department of Homeland Security
(Homeland Security Television 2011)

Most of the Internet infrastructure, including communication systems and platforms, are privately owned, yet it is largely in the hands of state authorities to act upon its security. Public-private partnerships (PPPs), a form of cooperation between the state and the private sector, are widely seen as a necessity to combat terrorist use of the Internet. The fundamental character of PPP can be described as follows: "Its goal is to exploit synergies in the joint innovative use of resources and in the application of management knowledge, with optimal attainment of the goals of all parties involved, where these goals could not be attained to the same extent without the other parties" (Dunn-Cavelty and Suter 2009, 180).

Critical infrastructure protection against cyberterrorism is currently seen as an essential part of national security in numerous countries; a broad range of efforts are underway in the United States, Europe, and in other parts of the world in an attempt to better secure critical infrastructures. One of the key challenges for such protection efforts arises from the privatization and deregulation of many parts of the critical infrastructure, which creates a situation in which market forces alone are not sufficient to provide security in most of the critical infrastructure sectors. At the same time, the state is incapable of providing the full security on its own. Therefore, cooperation between the state and the private corporate sector in cyberprotection is not only useful but inevitable.

In reality, however, the relationship is often unequal, conducted on an ad hoc basis, and rarely formalized. Such cooperation runs into a number of challenges, including the borderless character of the Internet, different national laws related to terrorist use of the Internet, and limited knowledge of each other's expertise. PPPs have been defined as collaboration between a public sector (government) entity and a private sector (for-profit) entity to achieve a specific goal or set of objectives. These partnerships have been discussed in narrow ways in the scholarly literature with regard to homeland security (such as emergency management or critical infrastructure protection). However, as Busch and Givens (2012) noted, "The scholarly literature has not yet caught up to the practitioner understanding of public-private partnerships' prominence in homeland security." The public-private partnerships hold great promise, but also face significant obstacles that will need to be overcome.

PPPs offer numerous advantages. A recent hacking incident may highlight the advantages of the PPP model for cybersecurity. In June 2011, Google publicly disclosed that individuals in China illegally accessed the personal email accounts of several senior US government officials. This was allegedly done by "phishing," a method of fraudulently obtaining a user's information through fabricated emails asking for usernames, passwords, and related data. Google notified the FBI about the incident. The White House National Security Council, as well as the Department of Homeland Security (DHS), followed up with Google to assess the incident's impact. An understanding of this attack's sources and methods provided greater knowledge of cybersecurity threats to public and private sector organizations. As this incident demonstrates, PPPs are critical to effective cybersecurity. The advantages, as detailed by Busch and Givens (2012) in their online article, include the following:

- *Hiring:* The private sector helps the public sector fill personnel needs more effectively than the government acting independently. Private sectors are much faster than the government in hiring people—they do not have lags from security clearances or background checks. This, in turn, creates value for the public sector. They work shoulder-to-shoulder with government counterparts in public sector homeland security offices. As a result, the homeland security workforce benefits from the hiring speed of the private sector.
- *Resources:* By orienting resources toward homeland security applications, businesses, government, and the public can benefit. Government gains from privately produced products and services, firms' sales of cybersecurity products increase, and public safety is enhanced. Private companies forge an advantageous triangular relationship among these stakeholders by using their resources for homeland security purposes.
- *Specialization:* By participating in homeland security activities, private sector actors develop specializations in functional areas (enhancing public sector performance), thus allowing the public sector to devote personnel and resources to other critical activities.
- *Building trust:* Over time, repeated interaction and collaboration may actually build trust across the government-business divide. Whether developing plans for the future, or responding to an emergency, trust is invaluable in fostering effective, mutually beneficial outcomes.
- *Promoting innovations:* Public-private partnerships can also serve as catalysts for new technological innovations. For example, two growing DHS initiatives promote private sector innovation for homeland security-related challenges: the System Efficacy through Commercialization, Utilization, Relevance and Evaluation (SECURE) program, and its sister program, FutureTECH. This departs from the traditional model of government-funded research and development, in that DHS provides clear requirements and design specifications to prospective vendors via public announcements.

Indeed, information technology firms' partnership with public sector agencies is essential in protecting national cybersecurity and counterterrorism objectives. Several well-known companies routinely partner with the government to share information and address cyberterrorism challenges. For example, the National Cyber Security Alliance (NCSA) is an organization that raises awareness about cybersecurity issues and empowers computer users to protect themselves against electronic threats. PPPs are critical to the NCSA mission and functions. The NCSA board includes representatives from numerous national firms, including AT&T Services, Inc., Cisco Systems, Lockheed Martin, Microsoft, Google, Facebook, Bank of America, SAIC, and Visa. Demonstrating linkages between the NCSA and federal government, the White House and the DHS promoted the most visible NCSA initiative, National Cyber Security Awareness Month, in 2010 (Busch and Givens 2012).

Nevertheless, there are instances in which the unique attributes of the two sectors make PPPs limited or inappropriate. As described above, PPPs by definition require complementary goals, mutual trust, clear goals and strategies, clear distribution of responsibilities, and similar task-oriented thinking. But there are significant differences across the sectors in terms of mentality, agenda, priorities, and methods. It has become clear that when it comes to critical infrastructure protection, the interests of the private sector and that of the state are only partially convergent and that therefore synergy effects are not always easily obtained. Specifically, private companies fear that sensitive information that is passed on to the state might not be treated with the necessary degree of confidentiality and might cause damage to their interests. Moreover, the private sector perceives the issue mainly from the perspective of business administration and regards it primarily as a matter of ensuring profits and continuity.

The need to mediate between the private and the public sectors is crucial. Currently, each private company, Internet provider, or online platform (e.g., Google, Yahoo!, Facebook, Twitter) interacts separately and independently with various government agencies (e.g., CIA, NSA, DHS, TSA, FBI, Department of Defense). Figure 11.1 presents the resulting PPP mode. Multiple agencies, companies, and organizations in both sectors are shown to be collaborating inefficiently. Instead of working together collaboratively, the agencies and companies make specific agreements with one other counterpart in the opposite sector at a time, instead of many at once.

A public–private sector collaboration mechanism (public-private mediating mechanism, or PPM) could alter this diagram significantly. A PPM is an organization whose role is to negotiate or determine how best to allow multiple agencies and organizations from both sectors to collaborate together at the same time. It eliminates the necessity for the tedious individual agreements that overlap or avoid other agencies that could improve a collective counterterrorism atmosphere. If a mediating organ such as the PPM is introduced, as shown in figure 11.2, everything looks much simpler and much

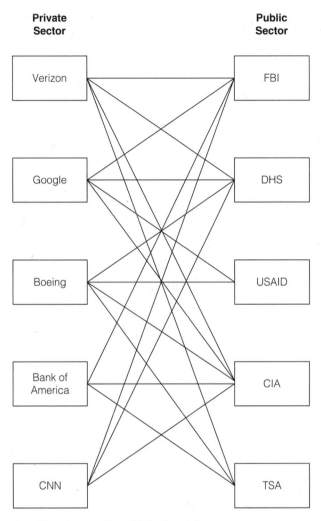

Figure 11.1. The Nonmediated PPP Model

more efficient. Although a PPM has yet to be created, the second diagram illustrates how much more effective and succinct a multiple-collaborative method organized by a PPM would be.

Finally, as often noted, a serious problem is that the majority of existing instances of PPPs are too narrow. Most Internet security PPPs involve cooperation between a specialized agency and selected private sector partners

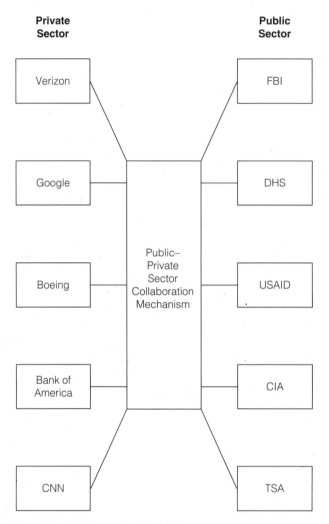

Figure 11.2.   The Mediated PPP Model

(e.g., infrastructure operators, Internet companies). This cooperation design does not allow for the needed horizontal (private to public) and vertical (within the private sector or within the public sector) integration of public and government agencies on the one hand and private companies on the other hand. One case study illustrates how these differences may harm the potential cooperation: In 2008, teams of government scientists identified a cybervulnerability in the US Bulk Power System, drafted a list of remedies to address the vulnerability, distributed the list to electrical companies, and provided a timeline for implementation (Busch and Givens 2012). Despite these proactive steps and the apparent mutual interest in addressing these vulnerabilities, in reality the private sector complied only minimally with these recommendations.[5] This example shows differences between public and private sector approaches to cybersecurity. Furthermore, there is a dissonance that is hard to overcome between the logic of security and the logic of PPPs: a core task of the state is generating security for its citizens; therefore, it is an extremely delicate matter for the government to pass on its responsibility in this area to the private sector (Dunn-Cavelty and Suter 2009).

Thus, despite the accepted importance of public-private sector cooperation on cybersecurity initiatives, actual cooperation may be less common than one imagines. In light of the challenges that PPPs face, the public and private sectors would be well served by showing why government-business partnerships are necessary, and how their existence benefits homeland security. PPPs are easy when both government and business immediately benefit. As cybersecurity PPPs continue to grow, there is also a compelling need for scholarly and academic involvement. Academics can make valuable contributions in studying actions, initiatives, and special projects that add value to PPPs in homeland security in general and in cyberspace in particular. Researchers can work with industry leaders to enhance tools to share best practices. For example, the need for a mediating frame between the private and the public agencies could provide an excellent starting point for this cross-sector collaboration. Scholars are in an ideal position to make analytical connections and provide a theoretical framework for successful PPP management (Busch and Givens 2012).

* * *

Terrorist use of the Internet continues to expand and proliferate, change and evolve. This book reviews the current changes and the emerging trends, thus updating the path-breaking work that *Terror on the Internet* presented in 2006. Comparing the findings from the two studies reveal that terrorists' presence and the use of cyberspace is today more sophisticated, richer,

---

[5] See Hon. John D. Dingell, "Protecting the Electrical Grid from Cybersecurity Threats," testimony before the Subcommittee on Energy and Air Quality, of the Committee on Energy and Commerce, US House of Representatives, 110th Cong., 2nd Sess. (September 11, 2008), 128.

and broader than a decade ago. New online platforms are emerging and quickly being adapted by savvy terrorists, and new cyberthreats are looming. Counterterrorism agencies in many countries have responded to these new challenges, but the responses often have raised concerns about the prices we pay in terms of civil liberties, privacy, freedom of expression, and more. We live in a dangerous world, threatened by terrorism, and intelligence agencies should do their utmost to protect us against terrorist plots. But they must also do it intelligently, ethically, and with minimal harm to our democratic values.

References

304th MI BN OSINT Team. 2008. "Al Qaida-Like Mobile Discussions & Potential Creative Uses." Intelligence Resource Program, Federation of American Scientists. Accessed February 12, 2014. http://fas.org/irp/eprint/mobile.pdf.

Adnkronos International. 2008. "Egypt: Sunni Scholars Sanction 'Electronic Jihad.'" October 16. http://www1.adnkronos.com/AKI/English/Security/?id=3.0.2595019598.

The Advertiser. 2008. "'Facebook' Terrorism Investigation." *The Advertiser* (Adelaide, Australia), April 5.

Agence France-Presse. 2013. "11 Killed as Syria Rebels, Kurds Clash." Agence France-Presse, May 26. http://www.globalpost.com/dispatch/news/afp/130526/11-killed-syria-rebels-kurds-clash.

Ahlers, Mike M. 2004. "Blueprints for Terrorists?" CNN.com, October 19. http://www.cnn.com/2004/US/10/19/terror.nrc/.

Ahlqvist, Toni, Asta Bäck, Minna Halonen, and Sirkka Heinonen. 2008. *Social Media Road Maps: Exploring the Futures Triggered by Social Media.* VTT Tiedotteita—Valtion Teknillinen Tutkimuskeskus. VTT Research Notes 2454. http://www.vtt.fi/inf/pdf/tiedotteet/2008/T2454.pdf.

AIVD (General Intelligence and Security Service). 2012. *Jihadism on the Web: A Breeding Ground for Jihad in the Modern Age.* Amsterdam: Ministry of the Interior and Kingdom Relations of the Netherlands.

Akin, David. 2004. "Arrests Key Win for NSA Hackers." *Globe and Mail*, April 6: A16. http://www.theglobeandmail.com/servlet/ArticleNews/TPStory/LAC/20040406 /TERROR06/International/Idx/.

Al Raffie, Dina. 2012. "Whose Hearts and Minds? Narratives and Counter-Narratives of Salafi Jihadism." *Journal of Terrorism Research* 3 (2): 13–31.

Aly, Anne, Dana Weimann-Saks, and Gabriel Weimann. 2014. "Making 'Noise' Online: An Analysis of the 'Say No to Terror' Online Campaign." *Perspectives on Terrorism* 8 (5): 33–47. http://www.terrorismanalysts.com/pt/index.php/pot/article/view/376 /html.

Amble, John Curtis. 2012. "Combating Terrorism in the New Media Environment." *Studies in Conflict & Terrorism* 35 (5): 339–53.

American Civil Liberties Union (ACLU). 2003. "ACLU Says New Ashcroft Anti-Terror Proposal Undermines Checks and Balances." ACLU, February 7. http://www.aclu .org/SafeandFree/SafeandFree.cfm?ID=11803&c=206.

Anderson, Chris. 2006. *The Long Tail: Why the Future of Business Is Selling Less of More*. New York: Hyperion.

Anderson, Craig A., and Brad J. Bushman. 2001. "Effects of Violent Video Games on Aggressive Behavior, Aggressive Cognition, Aggressive Affect, Physiological Arousal and Prosocial Behaviour: A Meta-Analytic Review of the Scientific Literature." *Psychological Science* 12 (5): 353–59.

Anderson, Ross. 2008. *Security Engineering: A Guide to Building Dependable Distributed Systems*, 2nd ed. New York: Wiley.

Anti-Defamation League. 2002. *Jihad Online: Islamic Terrorists and the Internet*. New York: Anti-Defamation League. http://archive.adl.org/internet/jihad_online.pdf.

———. 2013. "Terry Lee Loewen Planned Airport Bombing for Al Qaeda." Extremism & Terrorism blog, December 13. http://blog.adl.org/extremism/terry-lee-loewen -airport-bomb-al-qaeda-revolution-muslim.

Archetti, Cristina. 2010. "Terrorism, Communication and the War of Ideas: Al-Qaida's Strategic Narrative as a Brand." Paper presented at the International Communication Association Annual Convention, Singapore, June 22–26.

———. 2012. *Understanding Terrorism in the Age of Global Media: A Communication Approach*. London: Palgrave Macmillan.

Armerding, Taylor. 2012. "Advanced Persistent Threats Can Be Beaten, Says Expert." CSO Online, August 6. http://www.csoonline.com/article/2131881/critical -infrastructure/advanced-persistent-threats-can-be-beaten--says-expert.html.

Arquilla, John, and David Ronfeldt. 2001. "The Advent of Netwar (revisited)." In *Networks and Netwars: The Future of Terror, Crime, and Militancy*, edited by John Arquilla and David Ronfeldt, 1–25. Santa Monica, CA: RAND Corporation.

———. 2003. "Networks, Netwars, and the Fight for the Future." *First Monday* 25: 1–25.

Arquilla, John, David Ronfeldt, and Michele Zanini. 2001. "Networks, Netwar and Information-Age Terrorism." In *Countering the New Terrorism*, edited by Ian O. Lesser, Bruce Hoffman, John Arquilla, David F. Ronfeldt, Michele Zanini, and Brian Michael Jenkins, 75–111. Santa Monica, CA: RAND Corporation.

Associated Press. 2002. "Using Terror as a Pretext." Wired.com, September 4. http:// archive.wired.com/politics/law/news/2002/09/54939.

———. 2011. "Hacking in the Netherlands Took Aim at Internet Giants." *New York Times*, September 5. http://www.nytimes.com/2011/09/06/technology/hacking-in -the-netherlands-broadens-in-scope.html.

———. 2012. "Terrorists May Be Eyeing Cyber Attacks on US." Fox News, March 7. http://www.foxnews.com/us/2012/03/07/terrorists-may-be-eyeing-cyber-attacks-on -us/.

———. 2013. "Number of Active Users at Facebook over the Years." Yahoo!News, May 1. http://news.yahoo.com/number-active-users-facebook-over-230449748.html.

Azuri, L. 2008. "Recent Rise in Sunni–Shi'ite Tension: Sunni–Shi'ite Hacker War on the Internet." MEMRI Inquiry & Analysis Series Report 480, December 16. http:// www.memri.org/report/en/0/0/0/0/0/0/3145.htm.

Ball, James. 2014. "NSA Collects Millions of Text Messages Daily in 'Untargeted' Global Sweep." *Guardian* (London), January 16. http://www.theguardian.com/world /2014/jan/16/nsa-collects-millions-text-messages-daily-untargeted-global-sweep.

Ball, James, Julian Borger, and Glenn Greenwald. 2013. "Revealed: How US and UK Spy Agencies Defeat Internet Privacy and Security." *Guardian* (London), September 5. http://www.theguardian.com/world/2013/sep/05/nsa-gchq-encryption-codes -security.

Bamford, James. 2012. "The NSA Is Building the Country's Biggest Spy Center." Wired. com, March 15. http://www.wired.com/threatlevel/2012/03/ff_nsadatacenter/all/.

Bar, Shmuel. 2006a. *Warrant for Terror: The Fatwas of Radical Islam and the Duty to Jihad*. Lanham, MD: Rowman and Littlefield Publishing.

———. 2006b. *Jihad Ideology in Light of Contemporary Fatwas*. Washington, DC: Hudson Institute.

BBC. 2009. "Islamists 'Promote Jihad in Jail.'" BBC News, November 15. http://news .bbc.co.uk/2/hi/uk_news/8361440.stm.

———. 2011. "Kosovan Admits Shooting US Airmen at Frankfurt Airport." BBC News, August 31. http://www.bbc.co.uk/news/world-europe-14727975.

———. 2013. "UK Cyber Defence Unit 'May Include Hackers.'" BBC News, October 22. http://www.bbc.com/news/technology-24613376.

Beaumont, Claudine. 2009. "Twitter Hacker by 'Iranian Cyber Army.'" *Daily Telegraph* (London), December 18. http://www.telegraph.co.uk/technology/twitter/6838993 /Twitter-hacked-by-Iranian-Cyber-Army.html.

Bender, Bryan. 2008. "Unwittingly Hosting Terror." *Boston Globe*, March 27. http:// www.boston.com/business/articles/2008/03/27/unwittingly_hosting_terror/.

Bergen, Peter. 2007. "War of Error." *New Republic*, October 15. http://www.newrepublic .com/article/politics/war-error.

———. 2013. "Are Mass Murderers Using Twitter as a Tool?" CNN.com, September 27. http://edition.cnn.com/2013/09/26/opinion/bergen-twitter-terrorism/.

Bergen, Peter, Bruce Hoffman, Michael Hurley, and Erroll G. Southers. 2013. *Jihadist Terrorism: A Threat Assessment*. Homeland Security Project. Washington, DC: Bipartisan Policy Center. http://bipartisanpolicy.org/sites/default/files/Jihadist%20 Terrorism-A%20Threat%20Assesment_0.pdf.

Bergin, Anthony, Sulastri Bte Osman, Carl Ungerer, and Nur Azlin Mohamed Yasin. 2009. *Countering Internet Radicalisation in Southeast Asia*. Australian Strategic Policy Institute, Special Report 22, March 6. https://www.aspi.org.au/publications /special-report-issue-22-countering-internet-radicalisation-in-southeast-asia.

Betz, David. 2008. "The Virtual Dimension of Contemporary Insurgency and Counterinsurgency." *Small Wars & Insurgencies* 19 (4): 510–40.

Blumenthal, Les. 2007. "U.S. Seeks to Counter Terrorists' Use of the Internet." *Kansas City Star*, September 23: A23.

Boccara, Marie-Hélène. 2004. "Islamist Websites and Their Hosts, Part I: Islamist Terror Organizations." MEMRI Special Report 31, July 16. http://memri.org/report /en/0/0/0/0/0/50/1174.htm.

Boccara, Marie-Hélène, and Alex Greenberg. 2004. "Islamist Websites and Their Hosts, Part II: Clerics." MEMRI Special Report 35, November 11. http://www.memri.org /report/en/0/0/0/0/0/50/1257.htm.

Bolz, Frank, Kenneth Dudonis, and David Schulz. 2002. *The Counterterrorism Handbook: Tactics, Procedures, and Techniques*. Boca Raton, FL: CRC Press.

Boyle, Louise, and Associated Press. 2013. "'Brother Bin Laden is a great inspiration to me': Authorities Thwart Suicide Bombing 'By Islamic Extremist Flight Technician Who Plotted to Blow Up Kansas Airport.'" *Daily Mail* (London), December 13. http: //www.dailymail.co.uk/news/article-2523387/Authorities-thwart-suicide-bombing -Islamic-extremist-flight-technician-plotted-blow-Kansas-airport.html.

Brahimi, Alia. 2010. *The Taliban's Evolving Ideology*. London School of Economics: Global Governance Working Paper.

Braun, Michael. 2008. "Drug Trafficking and Middle Eastern Terrorist Groups: A Growing Nexus?" Washington Institute for Near East Policy, July 25. http://www .washingtoninstitute.org/policy-analysis/view/drug-trafficking-and-middle-eastern -terrorist-groups-a-growing-nexus.

Brickey, Jonalan. 2012. "Defining Cyberterrorism: Capturing a Broad Range of Activities in Cyberspace." *CTC Sentinel* 5 (8): 4–6. https://www.ctc.usma.edu/posts /defining-cyberterrorism-capturing-a-broad-range-of-activities-in-cyberspace.

Buckley, Cara, and William K. Rashbaum. 2007. "4 Men Accused of Plot to Blow Up Kennedy Airport Terminals and Fuel Lines." *New York Times*, June 3. http://www .nytimes.com/2007/06/03/nyregion/03plot.html.

Bunt, Gary. 2000. *Virtually Islamic: Computer-Mediated Communication and Cyber Islamic Environment*. Cardiff: University of Wales Press.

———. 2003. *Islam in the Digital Age: E-Jihad, Online Fatwas and Cyber Islamic Environments*. London: Pluto.

———. 2009. *iMuslims: Rewiring the House of Islam*. Chapel Hill: University of North Carolina Press.

Busch, Nathan E., and Austen D. Givens. 2012. "Public-Private Partnerships in Homeland Security: Opportunities and Challenges." *Homeland Security Affairs* 8, Article 18 (October). http://www.hsaj.org/?article=8.1.18.

Carroll, Roy. 2012. "US Urged to Recruit Master Hackers to Wage Cyber War on America's Foes." *Guardian* (London), July 10. http://www.theguardian.com /technology/2012/jul/10/us-master-hackers-al-qaida.

Carter, Sara. 2013. "Feds Take Little Action Against U.S. Web Companies Hosting Sites Linked to Terror." *Washington Times*, May 15. http://www.washingtontimes.com /news/2013/may/15/feds-take-little-action-against-us-web-companies-h/.

Casebeer, William A., and James Russell. 2005. "Storytelling and Terrorism: Towards a Comprehensive 'Counter-Narrative Strategy.'" *Strategic Insights* 6 (3). http:// calhoun.nps.edu/bitstream/handle/10945/11132/casebeerMar05.pdf.

Chaffey, Dave. 2004. "E-marketing Strategy—An In-Depth Guide." Marketing Insights. http://www.marketing-insights.co.uk/wnim0402.htm.

Chaffey, Dave, Richard Mayer, Kevin Johnston, and Fiona Ellis-Chadwick. 2000. *Internet Marketing: Strategy, Implementation and Practice.* Harlow, UK: Person Education Ltd.

Change Institute. 2008. *Studies into Violent Radicalisation: The Beliefs, Ideologies and Narratives.* February. European Commission—Directorate-General for Justice, Freedom and Security. http://www.changeinstitute.co.uk/images/publications /changeinstitute_beliefsideologiesnarratives.pdf.

Cilluffo, Frank. 2013. "Countering Use of the Internet for Terrorist Purposes." Statement before the United Nations Security Council Counter Terrorism Committee, May 24, 2013. http://www.un.org/en/sc/ctc/docs/2013/2013-05-24_GWU-Session%20III.pdf.

Clapper, James R., Jr. 2014. "Statement for the Record: Worldwide Threat Assessment of the US Intelligence Community." Senate Select Committee on Intelligence, January 29. Office of the Director of National Intelligence. http://www.dni.gov/files /documents/Intelligence%20Reports/2014%20WWTA%20%20SFR_SSCI_29_Jan .pdf.

Clayton, Mark. 2013. "Terrorist Tweets: How Al Qaeda's Social Media Move Could Cause Problems." *Christian Science Monitor*, February 7. http://www.csmonitor .com/USA/2013/0207/Terrorist-tweets-how-Al-Qaeda-s-social-media-move-could -cause-problems.

Clean IT Project. 2012. "Clean IT Project—Detailed Recommendations Document." EDRi (European Digital Rights). August 28. http://www.edri.org/files/cleanIT _sept2012.pdf.

———. 2013. *Reducing Terrorist Use of the Internet.* January. http://www.nctv.nl /Images/reducing-terrorist-use-of-the-internet_tcm126-485438.pdf.

Cobb, Tyrus W. 2013. "Two Worrisome Scenarios on the Boston Bombers." National Security Forum, May 16. http://nationalsecurityforum.org/domestic-news/homeland -security/two-worrisome-scenarios-on-the-boston-bombers/.

Cole, Matthew. 2011. "New Al-Qaeda Video: American Muslims Should Buy Guns, Start Shooting People." ABC News, June 3. http://abcnews.go.com/Blotter/al-qaeda -video-buy-automatic-weapons-start-shooting/story?id=13704264.

Coll, Steve, and Susan Glasser. 2005. "Terrorists Turn to the Web as Base of Operations." *Washington Post*, August 7. http://www.washingtonpost.com/wp-dyn/content/article /2005/08/05/AR2005080501138.html.

Collin, Barry. 1997. "The Future of Cyberterrorism." *Crime & Justice International Journal* 13 (2). http://www.cjimagazine.com/archives/cji4c18.html?id=415.

Collins, Nick. 2010. "YouTube Radicalisation: Video Site Directed MP's Attacker to Extremist Videos." *Daily Telegraph* (London), November 4. http://www.telegraph .co.uk/news/uknews/terrorism-in-the-uk/8109613/YouTube-radicalisation-video -site-directed-MPs-attacker-to-extremist-videos.html.

———. 2013. "Cyber Terrorism Is 'Biggest Threat to Aircraft.'" *Daily Telegraph* (London), December 23. http://www.telegraph.co.uk/finance/newsbysector/transport /10526620/Cyber-terrorism-is-biggest-threat-to-aircraft.html.

Conway, Maura. 2002. "Reality Bytes: Cyberterrorism and Terrorist 'Use' of the Internet." *First Monday* 7. http://www.firstmonday.org/ojs/index.php/fm/issue/view /151.

———. 2005. "Terrorist 'Use' of the Internet and Fighting Back." Paper presented at the "Cybersafety: Safety and Security in a Networked World: Balancing Cyber-Rights and Responsibilities" conference, Oxford Internet Institute, Oxford University, UK, September 8–10. http://www.oii.ox.ac.uk/microsites/cybersafety/extensions /pdfs/papers/maura_conway.pdf.

Conway, Maura, and Lisa McInerney. 2008. "Jihadi Video & Auto-Radicalisation: Evidence from an Exploratory YouTube Study." In *Proceedings of the 1st European Conference on Intelligence and Security Informatics*, Esbjerg, Denmark, December 3–5. http://doras.dcu.ie/2253/2/youtube_2008.pdf.

Cooper, Abraham. 2009. *Facebook, YouTube +: How Social Media Outlets Impact Digital Terrorism and Hate*. Los Angeles: Simon Wiesenthal Center. http://www .wiesenthal.com/atf/cf/%7B54d385e6-f1b9-4e9f-8e94-890c3e6dd277%7D/NY -RELEASE.PDF.

COT/Transnational Terrorism, Security, and the Rule of Law (COT/TTSRL). 2007. *Lone-Wolf Terrorism*. Case Study for Work Package 3: Citizens and Governance in a Knowledge-Based Society. http://www.transnationalterrorism.eu/tekst/publications /Lone-Wolf%20Terrorism.pdf.

Coviello, Nicole E., Roderick J. Brodie, Richard W. Brookes, and Roger A. Palmer. 2003. "Assessing the Role of e-Marketing in Contemporary Marketing Practice." *Journal of Marketing Management* 19 (7–8): 857–81.

Crabbe, Nathan. 2011. "Al-Qaida Becoming Irrelevant, CNN Analyst Tells UF Audience." *Gainsville Sun*. October 12. http://www.gainesville.com/article/20111012 /ARTICLES/111019807.

Cronin, Audrey Kurth. 2006. "Cyber-Mobilization: The New *Levée en Masse*." *Parameters* (Summer): 77–87. http://strategicstudiesinstitute.army.mil/pubs /parameters/Articles/06summer/cronin.pdf.

Dal Cin, Sonya, Mark P. Zanna, and Geoffrey T. Fong. 2004. "Narrative Persuasion and Overcoming Resistance." In *Resistance and Persuasion*, edited by Eric S. Knowles and Jay A. Linn, 175–92. Mawah, NJ: Erlbaum.

Davis, Anthony. 2002. "The Afghan Files: Al-Qaeda Documents from Kabul." *Jane's Intelligence Review* 14 (2): 14–19.

Denning, Dorothy. 1998. *Information Warfare and Security*, New York: Addison-Wesley.

———. 2000. "Cyberterrorism: Testimony before the Special Oversight Panel on Terrorism." Statement to U.S. House of Representatives, Committee on Armed Services, May 23. http://www.stealth-iss.com/documents/pdf/CYBERTERRORISM .pdf.

———. 2001. "Activism, Hacktivism, and Cyberterrorism: The Internet as a Tool for Influencing Foreign Policy." In *Networks and Netwars: The Future of Terror, Crime, and Militancy*, edited by John Arquilla and David Ronfeldt, 239–88. Santa Monica, CA: RAND Corporation.

———. 2010. "Terror's Web: How the Internet Is Transforming Terrorism." In *Handbook of Internet Crime*, edited by Yvonne Jewkes and Majid Yar, 194–213. Cullompton, UK: Willan Publishing.

Department of Homeland Security. 2010. "DHS Terrorist Use of Social Networking Facebook Case Study." PublicIntelligence.net, December 5. http://publicintelligence .net/ufouoles-dhs-terrorist-use-of-social-networking-facebook-case-study/.

Department of Justice. 2010. "Pennsylvania Woman Indicted in Plot to Recruit Violent Jihadist Fighters and to Commit Murder Overseas." Office of Public Affairs. March 9. http://www.justice.gov/opa/pr/2010/March/10-ag-238.html.

Department of State. 2013. *Country Reports on Terrorism 2012.* Office of the Coordinator for Counterterrorism. May 2013. http://www.state.gov/documents/organization /210204.pdf.

Department of State. 2014a. "Foreign Terrorist Organizations." Accessed September 16, 2014. http://www.state.gov/j/ct/rls/other/des/123085.htm.

———. 2014b. "State Sponsors of Terrorism." Accessed September 16, 2014. http:// www.state.gov/j/ct/list/c14151.htm.

deYoung, Karen. 2006. "Spy Agencies Say Iraq War Hurting U.S. Terror Fight," *Washington Post*, September 24. http://www.washingtonpost.com/wp-dyn/content /article/2006/09/23/AR2006092301130.html.

Dobson, Christopher, and Ronald Paine. 1977. *The Carlos Complex: A Pattern of Violence.* London: Hodder and Stoughton.

Dodd, Vikram. 2010. "Roshonara Choudhry: I Wanted to Die ... I Wanted to Be a Martyr." *Guardian* (London), November 3. http://www.theguardian.com/uk/2010 /nov/04/stephen-timms-attack-roshonara-choudhry.

Dorgan, Bryan. 2013. "Cyber Terror Is the New Language of War." *Huffington Post*, July 17. http://www.huffingtonpost.com/sen-byron-dorgan/cyber-terror-is-the-new-l _b_3612888.html.

Dunn-Cavelty, Myriam, and Manuel Suter. 2009. "Public–Private Partnerships Are No Silver Bullet: An Expanded Governance Model for Critical Infrastructure Protection." *International Journal of Critical Infrastructure Protection* 2 (4): 179–87.

Electronic Frontier Foundation. 2014. "How the NSA's Domestic Spying Program Works." Accessed September 16, 2014. https://www.eff.org/nsa-spying/how-it -works.

Electronic Privacy Information Center. 2005. "The USA Patriot Act." May 24. Accessed April 2, 2014. http://www.epic.org/privacy/terrorism/usapatriot/.

Erwin, Michael. 2008. "Key Factors for the Recent Growth of the Afghan Insurgency." *CTC Sentinel* 1 (9): 9–11. http://www.ctc.usma.edu/posts/key-factors-for-the-recent -growth-of-the-afghan-insurgency.

Etzioni, Amitai. 1999. *The Limits of Privacy.* New York: Basic Books.

———. 2002. "Seeking Middle Ground on Privacy vs. Security." *Commentary*, October 15.

European Commission. 2008. "Radicalization Processes Leading to Acts of Terrorism." Expert Group on Violent Radicalization, European Commission. http://www .clingendael.nl/sites/default/files/20080500_cscp_report_vries.pdf.

Europol. 2012. *Annual Terrorism Situation and Trend Report.* European Police Office. https://www.europol.europa.eu/sites/default/files/publications/europoltsat.pdf.

Facebook. 2014. "Our Mission." Facebook.com. Accessed September 14, 2014. http:// newsroom.fb.com/company-info/.

Fahd, Shaykh Nasir bin Hamid al-. 2003. "A Treatise on the Legal Status of Using Weapons of Mass Destruction Against Infidels." May. Accessed April 9, 2014. http://ahlussunnahpublicaties.files.wordpress.com/2013/04/42288104-nasir-al-fahd -the-ruling-on-using-weapons-of-mass-destruction-against-the-infidels.pdf.

Farmer, Ben. 2013. "NATO Suffered 2,500 Cyber Attacks in 2012." *Daily Telegraph* (London), January 4. http://www.telegraph.co.uk/news/uknews/defence/10099525/Nato-suffered-2500-cyber-attacks-in-2012.html.

Federal Bureau of Investigation. 2013. "Botnets 101: What They Are and How to Avoid Them." June 5. http://www.fbi.gov/news/news_blog/botnets-101/botnets-101-what-they-are-and-how-to-avoid-them.

Finn, Peter. 2012. "Inspire, Al-Qaeda's English-language Magazine, Returns Without Editor Awlaki." *Washington Post*, May 2. http://www.washingtonpost.com/world/national-security/inspire-al-qaedas-english-language-magazine-returns-without-editor-awlaki/2012/05/02/gIQAiEPMxT_story.html.

Fink, Naureen Chowdhury, and Jack Barclay. 2013. Mastering the Narrative: Counterterrorism Strategic Communication and the United Nations. Washington, DC: Center on Global Counterterrorism Cooperation. http://www.globalcenter.org/wp-content/uploads/2013/03/Feb2013_CT_StratComm.pdf.

Forest, James J. F. 2012. "Perception Challenges Faced by Al-Qaeda on the Battlefield of Influence Warfare." *Perspectives on Terrorism* 6 (1): 8–22. http://www.terrorismanalysts.com/pt/index.php/pot/article/view/forest-perception-challenges.

Foster, Peter. 2013. "'Bogus' AP tweet about explosion at the White House wipes billions off US markets." *Daily Telegraph* (London), April 23. http://www.telegraph.co.uk/finance/markets/10013768/Bogus-AP-tweet-about-explosion-at-the-White-House-wipes-billions-off-US-markets.html.

Fouda, Yosri, and Nick Fielding. 2003. *Masterminds of Terror: The Truth Behind the Most Devastating Terrorist Attack the World Has Ever Seen.* New York: Arcade.

Freamon, Bernard K. 2003. "Martyrdom, Suicide, and the Islamic Law of War: A Short Legal History." *Fordham International Law Journal* 27: 299–369.

Freedman, Lawrence. 2006. *The Transformation of Strategic Affairs.* London: International Institute for Strategic Studies.

Friedman, Thomas. 2009. "America vs. The Narrative." *New York Times*, November 28. http://www.nytimes.com/2009/11/29/opinion/29friedman.html.

Friscolanti, Michael. 2013. "The Self-Radicalized Terrorist Next Door." *Maclean's*, July 12. http://www.macleans.ca/society/life/the-terrorist-next-door-2/.

Fuller, Graham. 2004. *The Future of Political Islam.* New York: Palgrave Macmillan.

Gadahn, Adam Yahiye. 1995. "Becoming Muslim." Free Republic. Accessed April 3, 2014. http://www.freerepublic.com/focus/f-news/1482394/posts.

Gardels, Nathan. 2009. "Mike McConnell: An American Spymaster on Cyberwar." *Huffington Post*, July 8. http://www.huffingtonpost.com/nathan-gardels/mike-mcconnell-an-america_b_227944.html.

Gardham, Duncan. 2010. "Anwar al-Awlaki: MI5 Warns of the al-Qaeda Preacher Targeting Britain." *Daily Telegraph* (London), June 11. http://www.telegraph.co.uk/news/uknews/terrorism-in-the-uk/7822761/Anwar-al-Awlaki-MI5-warns-of-the-al-Qaeda-preacher-targeting-Britain.html.

Gartenstein-Ross, David, and Laura Grossman. 2009. *Homegrown Terrorists in the U.S. and U.K.* Washington, DC: FDD Press.

Gerrig, Richard. 1993. *Experiencing Narrative Worlds.* New Haven, CT: Yale University Press.

Gertz, Bill. 2011. "Iran Militia Claims Credit for VOA Cyberstrike." *Washington Times*, February 22. http://www.washingtontimes.com/news/2011/feb/22/iran-militia-claims-credit-for-voa-cyberstrike/.

Gerwehr, Scott, and Sara Daly. 2006. "Al-Qaida: Terrorist Selection and Recruitment." In *The McGraw-Hill Homeland Security Handbook*, edited by David Kamien, 73–89. New York, McGraw-Hill.

Gillam, Carey. 2013. "Man Arrested for Suspected Suicide Plot to Blow Up Kansas Airport." Reuters.com, December 13. http://www.reuters.com/article/2013/12/13/us-usa-kansas-plot-idUSBRE9BC0UI20131213.

Glasser, Susan, and Steve Coll. 2005. "The Web as a Weapon." *Washington Post*, August 9. http://www.washingtonpost.com/wp-dyn/content/article/2005/08/08/AR2005080801018.html.

Goncalves, Jorge, Vassilis Kostakos, and Jayant Venkatanathan. 2013. "Narrowcasting in Social Media: Effects and Perceptions." In *Proceedings of the 2013 IEEE/ACM International Conference on Advances in Social Networks Analysis and Mining*, 502–9. New York: IEEE/ACM.

Goodenough, Patrick. 2013. "'Sexual Jihad' Fatwa Urges 'Temporary Marriages' for Syrian Rebels." CNS News, March 29. http://cnsnews.com/news/article/sexual-jihad-fatwa-urges-temporary-marriages-syrian-rebels.

Gorge, Mathieu. 2007. "Cyberterrorism: Hype or Reality?" *Computer Fraud & Security* 2007 (2): 9–12.

Gorman, Siobhan, and Julian E. Barnes. 2014. "Iranian Hacking to Test NSA Nominee Michael Rogers." *Wall Street Journal*, February 18. http://online.wsj.com/news/articles/SB10001424052702304899704579389402826681452.

Gorman, Siobhan, and Devlin Barrett. 2014. "White House Weighs Options for Revamping NSA Surveillance." *Wall Street Journal*, February 25. http://online.wsj.com/articles/SB10001424052702303880604579405640624409748.

Gorman, Siobhan, and Danny Yadron. 2013. "Iran Hacks Energy Firms, U.S. Says." *Wall Street Journal*, May 27. http://online.wsj.com/news/articles/SB10001424127887323336104578501601108021968.

Graham, Robert. 2001. "Carnivore FAQ (Frequently Asked Questions)." October 6. http://www.robertgraham.com/pubs/carnivore-faq.html. Archived at http://corz.org/public/docs/privacy/carnivore-faq.html.

Green, Melanie C., and Timothy C. Brock. 2000. "The Role of Transportation in the Pervasiveness of Public Narratives." *Journal of Personality and Social Psychology* 79 (5): 701–21.

Green, Melanie C., Jeffrey J. Strange, and Timothy C. Brock, ed. 2002. *Narrative Impact: Social and Cognitive Foundations*. Mahwah, NJ: Erlbaum.

Greenberg, Andy. 2010. "'Here You Have' Virus Writer Claims Credit in Video Denouncing Iraq Invasion." *Forbes*, September 13. http://www.forbes.com/sites/andygreenberg/2010/09/13/here-you-have-virus-writer-claims-credit-in-video-denouncing-iraq-invasion/.

Guttman, Nathan. 2012. "Should Terror Groups Be Able To Tweet and Use YouTube?" *Forward*, August 17: 1–8.

Hafner, Katie, and Saritha Rai. 2005. "Governments Tremble at Google's Bird's-Eye View." *New York Times*, December 20. http://www.nytimes.com/2005/12/20/technology/20image.html.

Halloran, Richard. 2007. "Strategic Communication." *Parameters* (Autumn): 4–14. http://strategicstudiesinstitute.army.mil/pubs/parameters/Articles/07autumn/halloran.pdf.

Halverson, Jeffrey, Steven Corman, and H. L. Goodall. 2011. *Master Narratives of Islamist Extremism*. New York: Palgrave Macmillan.

Hamilton College. 2002. Hamilton College Muslim America Poll. Zogby International /Arthur Levitt Public Affairs Center, May 30. http://www.hamilton.edu/news /MuslimAmerica/MuslimAmerica.pdf.

"Happy Snaps of a Suicide Bomber." 2004. *South African Star*, January 27. http://www .thestar.co.za/index.php?fSectionId=129&fArticleId=334925.

Harding, Thomas. 2007. "Terrorists 'Use Google Maps to Hit UK Troops.'" *Daily Telegraph* (London), January 13. http://www.telegraph.co.uk/news/worldnews /1539401/Terrorists-use-Google-maps-to-hit-UK-troops.html.

Harman, Jane. 2014. "Future Terrorists." *Los Angeles Times*, January 6. http://articles .latimes.com/2014/jan/06/opinion/la-oe-harman-terrorism-response-20140106

Hazan, D. 2008. "Women's Forums on Islamist Websites—Tools for Preparing Women to Carry Out Jihad and Suicide Operations." MEMRI Inquiry & Analysis Series Report 419, February 1. http://www.memri.org/report/en/0/0/0/0/0/0/2620.htm.

Heffelfinger, Christopher, 2010. "Waiting out the Islamist Winter: Creating and Effective Counter Narrative to 'Jihad.'" Paper presented at the GTReC ARC Linkage Project on Radicalisation Conference, November 8. http://artsonline.monash.edu.au /radicalisation/files/2013/03/conference-2010-counter-narratives-ch.pdf.

———. 2013. "The Risks Posed by Jihadist Hackers." *CTC Sentinel* 6 (71): 1–4. http://www.ctc.usma.edu/wp-content/uploads/2013/07/CTCSentinel-Vol6Iss71.pdf.

Hegghamer, Thomas. 2006. "Global Jihadism After the Iraq War." *Middle East Journal* 60 (1): 11–32.

Helfont, Samuel. 2009. *The Sunni Divide: Understanding Politics and Terrorism in the Arab Middle East*. Philadelphia: Center for Terrorism and Counterrorism at the Foreign Policy Research Institute.

Heller, Jeffrey. 2013. "Iran Ups Cyber Attacks on Israeli Computers: Netanyahu." Reuters, June 9. http://www.reuters.com/article/2013/06/09/us-israel-iran-cyber -idUSBRE95808H20130609.

Hershman, Tania. 2001. "Israel's 'First Internet Murder.'" Wired.com, January 19. Accessed April 15, 2014. http://archive.is/t3eR.

Hirsh, Michael. 2012. "Can Obama Safely Embrace Islamists?" *National Journal*, April 23. http://www.nationaljournal.com/blogs/decoded/2012/04/can-obama-safely -embrace-islamists--23.

Hoffman, Bruce. 2006a. *Inside Terrorism*, revised version. New York: Columbia University Press.

———. 2006b. "The Use of the Internet by Islamic Extremists," Testimony presented to the House Permanent Select Committee on Intelligence, May 4, 2006. http://www .au.af.mil/au/awc/awcgate/congress/hoffman_testimony4may06.pdf

Hoffman, Bruce, Mary Habeck, Aaron Y. Zelin, and Matthew Levitt. 2012. "Is al-Qaeda Central Still Relevant?" Washington Institute for Near East Policy. September 10. https://www.washingtoninstitute.org/policy-analysis/view/is-al-qaeda-central-still -relevant.

Holden, Michael. 2008. "Qaeda Videos Show Boys in Mock Attacks." Reuters, February 6. http://www.reuters.com/article/2008/02/06/us-iraq-idUSL1880448320080206.

Holtmann, Philipp. 2013. "Countering Al-Qaeda's Single Narrative." *Perspectives on Terrorism* 7 (2): 141–46. http://www.terrorismanalysts.com/pt/index.php/pot/article /view/262/html.

Home Office of the United Kingdom (Home Office). 2014. "Proscribed Terrorist Organisations." November 28. https://www.gov.uk/government/uploads/system /uploads/attachment_data/file/380939/ProscribedOrganisations.pdf.

Homeland Security Policy Institute/Critical Incident Analysis Group (HSPI/CIAG). 2007. *NETworked Radicalization: A Counter-Strategy.* Homeland Security Policy Institute (George Washington University) and Critical Incident Analysis Group (University of Virginia). http://homelandsecurity.gwu.edu/files/downloads/HSPI _Report_11.pdf.

Homeland Security Television. 2011. "Public-Private Partnerships for Homeland Security." YouTube video, 7:36. August 3. http://www.youtube.com /watch?v=ka6dgMxLrJI.

Horwitz, Sari. 2013. "Investigators Sharpen Focus on Wife of Dead Boston Bombing Suspect." *Washington Post*, May 3. http://www.washingtonpost.com/world/national -security/investigators-sharpen-focus-on-boston-bombing-suspects-widow/2013/05 /03/a2cd9d28-b413-11e2-baf7-5bc2a9dc6f44_story.html.

Hudson, John. 2012. "U.S. Counter-Terrorism Hackers Fight Al Qaeda One Prank at a Time." The Wire, May 24. http://www.thewire.com/global/2012/05/us-counter -terrorism-hackers-fight-al-qaeda-one-prank-time/52747/.

Ibrahim, Raymond. 2013. "Fatwa Permits Rape of Non-Sunni Women in Syria." FrontpageMag.com, April 4. http://www.frontpagemag.com/2013/raymond-ibrahim /fatwa-permits-rape-of-non-sunni-women-in-syria/.

Intelligence and Terrorism Information Center. 2007. "The Main Points of Ayman al-Zawahiri's Statement as Broadcasted on Al-Jazeera TV." Israel Intelligence Heritage and Commemoration Center, March 11. http://www.terrorism-info.org.il/data/pdf /PDF_07_058_2.pdf.

Interactive Advertising Bureau. 2013. "IAB Internet Advertising Revenue Report." PwC and Interactive Advertising Bureau. April. Accessed April 12, 2014. http://www.iab .net/media/file/IAB_Internet_Advertising_Revenue_Report_FY_2012_rev.pdf.

Internet Haganah. 2011. "Jose-Pimentel, a-k-a Mohammad-Yusuf." November 21. Accessed March 4, 2014. http://forum.internet-haganah.com/archive/index.php/t -358.html.

The Investigative Project on Terrorism. 2014. "State of New York v. Pimentel, Jose." Washington, DC. Accessed September 18, 2014. http://www.investigativeproject.org /case/602.

Jacobson, Michael. 2008. "Why Terrorists Quit: Gaining from Al Qa'ida's Losses." *CTC Sentinel* 1 (8): 1–4. https://www.ctc.usma.edu/posts/why-terrorists-quit-gaining -from-al-qaida%E2%80%99s-losses.

———. 2010a. *Learning Counter-Narrative Lessons from Cases of Terrorist Dropouts.* The Hague: National Coordinator for Counterterrorism. https://www .washingtoninstitute.org/uploads/Documents/opeds/4b7aaf56ca52e.pdf.

———. 2010b. "Terrorist Financing and the Internet." *Studies in Conflict & Terrorism* 33 (4): 353–63.

Jenkins, Brian. 1975. *International Terrorism.* Los Angeles: Crescent Publication.

———. 2011. *Is Al Qaeda's Internet Strategy Working?* Santa Monica, CA: RAND Corporation. http://www.rand.org/pubs/testimonies/CT371.html

Jihadi Websites Monitoring Group. 2010. "Periodical Review: Fatwas—October 2010." November. Herzliya, Israel: International Institute for Counter-Terrorism. http://www .ict.org.il/Article.aspx?ID=356.

"Jihadist Forum Suggests YouTube Invasion." 2008. *Daily Telegraph* (London), December 4. http://www.telegraph.co.uk/news/worldnews/northamerica/usa /3547072/Jihadist-forum-calls-for-YouTube-Invasion.html.

Johnston, Philip. 2007. "MI5: Al-Qa'eda Recruiting UK Children for Terror." *Daily Telegraph* (London), November 5. http://www.telegraph.co.uk/news/uknews /1568363/MI5-Al-Qaeda-recruiting-UK-children-for-terror.html.

Joint Chiefs of Staff. 2010. *Psychological Operations* (Joint Pub. 3-13.2). January 7. Washington, DC: US Government Printing Office.

Jost, Janis, and Curti Covi. 2013. *Terrorismus in London und Paris? Zwischen Ideologie und Frustration* [Terrorism in London and Paris? Between ideology and frustration]. *ISPK Policy Brief* No. 2, June 18. http://www.ispk.uni-kiel.de/fileadmin/user_upload /Kieler%20Analysen%20zur%20Sicherheitspolitik/ISPK_Policy_Brief/ISPK_PB2 __18.06.2013_.pdf.

Kaplan, Eben. 2009. "Terrorists and the Internet." Washington, DC: Council on Foreign Relations. http://www.cfr.org/terrorism-and-technology/terrorists-internet/p10005.

Kaplow, Louis, and Steven Shavell. 2002. *Fairness versus Welfare*. Cambridge, MA: Harvard University Press.

Kassimeris, George, and John Buckley. 2010. *The Ashgate Research Companion to Modern Warfare*. Farnham, UK: Ashgate.

Katz, Mark. 2011. "The 'War on Terror': Future Directions." Middle East Policy Council. January 25. http://www.mepc.org/articles-commentary/commentary/war-terror -future-directions.

Katz, Rita, and Josh Devon. 2014. "Jihad on YouTube." SITE Special Report, January 15. https://news.siteintelgroup.com/Featured-Article/jihad-on-youtube.html.

Katz, Rita, and Adam Raisman. 2013. "Special Report on the Power Struggle Between al-Qaeda Branches and Leadership: Al-Qaeda in Iraq vs. Al-Nusra Front and Zawahiri." SITE Special Report, June. http://news.siteintelgroup.com/index.php/18 -articles-a-analysis/3195-special-report-on-the-power-struggle-between-al-qaeda -branches-and-leadership-al-qaeda-in-iraq-vs-al-nusra-front-and-zawahiri.

———. 2014. "Syrian Jihad: The Weakening of al-Qaeda's Leadership." SITE Monitoring Service, April 8. https://news.siteintelgroup.com/Articles-Analysis /syrian-jihad-the-weakening-of-al-qaeda-s-leadership.html.

Kennedy, Jonathan and Gabriel Weimann. 2011. "The Strength of Weak Terrorist Ties." *Terrorism and Political Violence* 23 (2): 201–12.

Khatchadourian, Raffi. 2007. "Azzam the American: The Making of an Al Qaeda Homegrown." *New Yorker*, January 22. http://www.newyorker.com/reporting/2007 /01/22/070122fa_fact_khatchadourian.

Khayat, M. 2013. "Jihadis' Responses to Widespread Decline In Participation On Jihadi Forums; Increased Use Of Twitter." MEMRI Inquiry & Analysis Series Report 955, March 29. http://www.memri.org/report/en/0/0/0/0/0/0/7106.htm.

King, Meg. 2014. "Opinion: The Errors Between 1s and 0s." CNN Security Clearance, January 17. http://security.blogs.cnn.com/2014/01/17/opinion-the-errors-between-1s -and-0s/.

Klein, Ezra, and Evan Soltas. 2014. "Wonkbook: How Americans Feel About Inequality." *Washington Post* blog, January 21. http://www.washingtonpost.com/blogs/wonkblog /wp/2014/01/21/wonkbook-how-americans-feel-about-inequality/.

Kohlmann, Evan. 2006. "The Real Online Terrorist Threat." *Foreign Affairs*. September /October. http://www.foreignaffairs.com/articles/61924/evan-f-kohlmann/the-real -online-terrorist-threat.

———. 2009. "A Web of Lone Wolves." *Foreign Policy*, November 13. http://www .foreignpolicy.com/articles/2009/11/13/a_web_of_lone_wolves.

———. 2011. "The Antisocial Network: Countering the Use of Online Social Networking Technologies by Foreign Terrorist Organizations." Testimony before the House Committee on Homeland Security on "Jihadist Use of Social Media— How to Prevent Terrorism and Preserve Innovation," December 6. http://homeland. house.gov/sites/homeland.house.gov/files/Testimony%20Kohlmann%5B1%5D.pdf.

Kovacs, Eduard. 2013a. "US Department of State, Pentagon Websites Hacked by Tunisian Cyber Army." Softpedia News, January 3. http://news.softpedia.com/news /Week-4-of-Operation-Ababil-2-Hackers-to-Attack-9-US-Banks-318056.shtml.

———. 2013b. "Week 4 of Operation Ababil 2: Hackers to Attack 9 US Banks." Softpedia News, January 3. http://news.softpedia.com/news/Week-4-of-Operation -Ababil-2-Hackers-to-Attack-9-US-Banks-318056.shtml.

Kull, Steven, Clay Ramsay, Stephen Weber, Evan Lewis, and Ebrahim Mohseni. 2009. *Public Opinion in the Islamic World on Terrorism, al Qaeda, and US Policies*. WorldPublicOpinion.org. Program on International Policy Attitudes, University of Maryland. http://www.worldpublicopinion.org/pipa/pdf/feb09/STARTII_Feb09_rpt .pdf.

Kumar, Mohit. 2012. "US Authorities: Iranian Hackers Are Becoming a Real Pain." TheHackerNews.com, October 14. http://thehackernews.com/2012/10/us-authorities -iranian-hackers-are.html/.

Kumar, Nirmalya. 1999. "Internet Distribution Strategies: Dilemmas for the Incumbent." *Financial Times*, March 15: 6–7. http://faculty.london.edu/nkumar/assets/documents /Internet_distribution_strategies_-_Dilemmas_for_the_incumbent.pdf.

Labi, Nadya. 2006. "Jihad 2.0." *Atlantic Monthly*, July 1. http://www.theatlantic.com /doc/prem/200607/online-jihad.

Lal, Vinay. 2002. "Terror and Its Networks: Disappearing Trails in Cyberspace (draft)." The Nautilus Institute. http://oldsite.nautilus.org/gps/virtual-diasporas/paper/Lal .html.

Landler, Mark, and John Markoff. 2007. "Digital Fears Emerge After Data Siege in Estonia." *New York Times*, May 29. http://www.nytimes.com/2007/05/29/technology /29estonia.html.

Langford, Duncan. 1998. "Ethics @ the Internet: Bilateral Procedures in Electronic Communication." In *Cyberspace Divide: Equality, Agency and Policy in the Information Society*, edited by Brian D. Loader, 98–112. London: Routledge.

Lappin, Yaakov. 2014. "3 East Jerusalem al-Qaida Recruits Arrested, 'Planned Massive Bombings.'" *Jerusalem Post*, January 22. http://www.jpost.com/Defense/3-al-Qaida -recruits-arrested-planned-massive-bombings-339002.

Laub, Zachary. 2014. "Backgrounder: Hamas." Council on Foreign Relations. Accessed January 30, 2014. http://www.cfr.org/israel/hamas/p8968.

Lentini, Peter. 2013. *Neojihadism: Towards a New Understanding of Terrorism and Extremism?* Cheltenham, UK: Edward Elgar Publishing.

Leuprecht, Christian, Todd Hataley, Sophia Moskalenko, and Clark McCauley. 2009. "Winning the Battle but Losing the War? Narrative and Counter-Narratives Strategy."

Perspectives on Terrorism 3 (2): 25–35. http://www.terrorismanalysts.com/pt/index .php/pot/article/view/68/html.

———. 2010. "Containing the Narrative: Strategy and Tactics in Countering the Storyline of Global Jihad." *Journal of Policing, Intelligence and Counter Terrorism* 5 (1): 42–57.

Levitt, Matthew. 2009. "Radicalization: Made in the USA?" HSPI Commentary Series, June 2. Washington, DC: Homeland Security Policy Institute. http://homelandsecurity. gwu.edu/sites/homelandsecurity.gwu.edu/files/downloads/Commentary_3_HSPI .pdf.

Lewis, Bernard. 1988. *The Political Language of Islam*. Chicago: University of Chicago Press.

Lewis, James A. 2002. *Assessing the Risks of Cyber Terrorism, Cyber War and Other Cyber Threats*. Washington, DC: Center for Strategic and International Studies. December. http://csis.org/files/media/csis/pubs/021101_risks_of_cyberterror.pdf.

Lewis, Peter. 2001. "The Tools of Freedom and Security." *Fortune*, October 29: 200–4.

Lia, Brynjar, 2008. "Al-Qaida's Appeal: Understanding Its Unique Selling Points." *Perspectives on Terrorism* (2) 8: 3–10. http://www.terrorismanalysts.com/pt/index .php/pot/article/view/44/html.

Lieberman, Joseph I., and Susan M. Collins. 2011. *A Ticking Time Bomb: Counterterrorism Lessons from the U.S. Government's Failure to Prevent the Fort Hood Attack*. US Senate Committee on Homeland Security and Governmental Affairs, February 3. http://www.hsgac.senate.gov//imo/media/doc/Fort_Hood/FortHoodReport.pdf

Londono, Ernesto. 2011. "U.S. Military, Taliban Use Twitter to Wage War." *Washington Post*, December 18. http://articles.washingtonpost.com/2011-12-18/world/35284991 _1_isafmedia-abalkhi-social-media.

Ludlow, Peter. 2013. "What Is a 'Hacktivist'?" *New York Times* Opinionator blog, January 13. http://opinionator.blogs.nytimes.com/2013/01/13/what-is-a-hacktivist/.

Lungu, Angela Maria. 2001. "War.com: The Internet and Psychological Operations." *Joint Force Quarterly* 28: 14–17. http://www.dtic.mil/doctrine/jfq/jfq-28.pdf.

Lynch, Marc. 2006. "Al-Qaeda's Media Strategies." *The National Interest*, Spring. http://nationalinterest.org/article/al-qaedas-media-strategies-883.

Lyon, David. 2001. *Surveillance Society: Monitoring Everyday Life*. London: Open University Press.

Macdonald, Alistair. 2011. "U.K. Detects 'Talk' About Internet Terror Attack." *Wall Street Journal*, November 25. http://online.wsj.com/article/SB10001424052970203 764804577060272839781902.html.

Malik, Kenan. 2009. *From Fatwa to Jihad: The Rushdie Affair and Its Legacy*. London: Atlantic.

Mantel, Barbara. 2009. "Terrorism and the Internet." *CQ Global Researcher*, November 1. http://cqresearcherblog.blogspot.com/2009/11/should-governments-block-terrorist -web.html.

Marcus, Itamar, and Nan Jacques Zilberdik. 2012. "Mother Places Suicide Belt on Child, on Fatah Facebook Page in Lebanon." Palestinian Media Watch, October 29. http:// www.palwatch.org/main.aspx?fi=157&doc_id=7704.

Mazzocco, Philip J., and Melanie C. Green. 2011. "Narrative Persuasion in Legal Settings: What's the Story?" *The Jury Expert* 23 (3): 27–38. http://www.thejuryexpert .com/2011/05/narrative-persuasion/.

McCants, William. 2011. "Testimony: Jihadist Use of Social Media—How to Prevent Terrorism and Preserve Innovation." US House of Representatives: Subcommittee on Counterterrorism and Intelligence, December 6. http://homeland.house.gov/sites /homeland.house.gov/files/Testimony%20McCants.pdf.

McCarthy, Tom. 2014. "Obama Announces New Limits on NSA Surveillance Programs—Live Reaction." *Guardian* (London), January 17. http://www.theguardian .com/world/2014/jan/17/obama-nsa-surveillance-reforms-speech-live.

McCullagh, Declan. 2003. "Military Is Worried about Web Leaks." CNET News, January 16. http://news.cnet.com/2100-1023-981057.html.

MEMRI Jihad & Terrorism Studies Project. 2001. "Debating the Religious, Political and Moral Legitimacy of Suicide Bombings Part 1: The Debate over Religious Legitimacy." MEMRI Inquiry & Analysis Series Report 53, May 3. http://www .memri.org/report/en/0/0/0/0/0/0/451.htm.

———. 2004. "Reactions to Sheik Al-Qaradhawi's Fatwa Calling for the Abduction and Killing of American Civilians in Iraq." Special Dispatch Series 794. October 6. http://www.memri.org/report/en/0/0/0/0/0/0/1231.

———. 2013a. "Fatwa on Minbar Al-Tawhid Wal-Jihad (MTJ) Discusses Permissibility of Bombing European Synagogues, Churches." Special Dispatch Series 5272. April 15. http://www.memri.org/report/en/0/0/0/0/0/0/7131.htm.

———. 2013b. "Jihad and Terrorism Threat Monitor (JTTM) Weekend Summary." Special Announcement No. 261. October 12. http://www.memri.org/report/en/0/0/0 /0/0/50/7457.htm.

MEMRI Jihad & Terrorism Threat Monitor. 2010. "Fatwas Posted on the Website of Abu Muhammad Al-Maqdisi Permit Targeting Infidel Companies Such As Coca-Cola and McDonald's—and Kidnapping and Killing Tourists in Muslim Countries." Special Dispatch Series 3390, November 19. http://www.memrijttm.org/content/en /report.htm?report=4779.

———. 2011a. "New Al-Qaeda Al-Sahab Video Reiterates: Carry Out One-Man Jihad Operations in the West" Special Dispatch Series 3885, June 3. http://www.memrijttm .org/new-al-qaeda-al-sahab-video-reiterates-carry-out-one-man-jihad-operations-in -the-west.html.

———. 2011b. "New Al-Qaeda Magazine for Women Encourages Them to Fulfill Their Role in Supporting Jihad and the Mujahideen." Special Dispatch Series 3651, March 8. http://www.memrijttm.org/new-al-qaeda-magazine-for-women-encourages-them -to-fulfill-their-role-in-supporting-jihad-and-the-mujahideen.html.

———. 2012. "Al-Qaeda's E-Magazine for Women Encourages Them to Disseminate Ideology of Jihad and Martyrdom." Special Dispatch Series 4500, February 15. http://www.memrijttm.org/al-qaedas-e-magazine-for-women-encourages-them-to -disseminate-ideology-of-jihad-and-martyrdom.html.

———. 2013a. "Fatwas for the Land of Jihad—Part II: Young Men from the Middle East, Europe and the U.S. Turn to Minbar Al-Tawhid Wal-Jihad Clerics for Guidance Before Departing for Syria." December 17. http://www.memrijttm.org/fatwas-for-the -land-of-jihad-part-ii-young-men-from-the-middle-east-europe-and-the-us-turn-to -minbar-al-tawhid-wal-jihad-clerics-for-guidance-before-departing-to-syria.html.

———. 2013b. "Taliban Fatwa Threatens Journalists, Television Hosts, Analysts and Other Media Personalities in Pakistan with Death." November 4. http://www .memrijttm.org/taliban-fatwa-threatens-journalists-television-hosts-analysts-and -other-media-personalities-in-pakistan-with-death.html.

Meo, Nick, Ruth Sherlock, and Carol Malouf. 2012. "Hizbollah Debates Dropping Support for the Regime of President Bashar al-Assad." *Daily Telegraph* (London), October 27. http://www.telegraph.co.uk/news/worldnews/middleeast/lebanon /9638058/Hizbollah-debates-dropping-support-for-the-regime-of-President-Bashar -al-Assad.html.

Merriam, Lisa. 2011. "The Al-Qaeda Brand Died Last Week." *Forbes*, October 6. http://www.forbes.com/sites/realspin/2011/10/06/the-al-qaeda-brand-died-last-week/.

Michael, George, and Kassem M. Wahba, trans. and ed. 2001. "Text: Bin Laden Discusses Attacks on Tape." *Washington Post*, December 13. http://www .washingtonpost.com/wp-srv/nation/specials/attacked/transcripts/binladentext _121301.html.

Middle East Forum. 2004. "The Qaradawi Fatwas." *The Middle East Quarterly* 9 (3): 78–80. http://www.meforum.org/646/the-qaradawi-fatwas.

Middle East Times. 2008. "Cyber Terrorism: Perils of the Internet's Social Networks." *Middle East Times*, September 8.

MideastWeb. 2013. "Osama Bin Laden's 1998 Fatwa." MideastWeb.com. Accessed December 3, 2013. http://www.mideastweb.org/Osamabinladen2.htm.

Minei, Elizabeth, and Jonathan Matusitz. 2011. "Cyberterrorist Messages and Their Effects on Targets: A Qualitative Analysis." *Journal of Human Behavior in the Social Environment* 21 (8): 995–1019.

———. 2012. "Cyberspace as a New Arena for Terroristic Propaganda: An Updated Examination." *Poiesis & Praxis* 9 (1–2): 163–76.

Miniwatts Marketing Group. 2013. "Internet World Stats: Usage and Population Statistics." Internet World Stats. Accessed May 14, 2014. http://www .internetworldstats.com/.

Morgan, Nigel, Graham Jones, and Ant Hodges. 2012. *Social Media: The Complete Guide to Social Media From The Social Media Guys*. http://rucreativebloggingfa13 .files.wordpress.com/2013/09/completeguidetosocialmedia.pdf

Morris, Loveday. 2010. "The Anatomy of a Suicide Bomber." *National* (Abu Dhabi), October 24. http://www.thenational.ae/news/the-anatomy-of-a-suicide-bomber.

Mowatt-Larssen, Rolf. 2010. "Al Qaeda's Nuclear Ambitions." *Foreign Policy*, November 16. http://www.foreignpolicy.com/articles/2010/11/16/al_qaedas_nuclear _ambitions.

Mozes, Tomer, and Gabriel Weimann. 2010. "The E-Marketing Strategy of Hamas." *Studies in Conflict & Terrorism* 33 (3): 211–25.

Mueller, John, and Mark Stewart. 2011. "Terror, Security, and Money: Balancing the Risks, Benefits, and Costs of Homeland Security." Paper presented at the "Terror and the Economy: Which Institutions Help Mitigate the Damage?" panel at the Annual Convention of the Midwest Political Science Association, Chicago, IL, April 1. http://politicalscience.osu.edu/faculty/jmueller//MID11TSM.PDF

Mulrine, Anna. 2013. "How an 8-by-10 Foot Plywood Table Holds the Future of America's National Security." *Business Insider*, September 16. http://www .businessinsider.com/the-military-is-recruiting-middle-school-hackers-2013-9.

Nacos, Brigitte. 2002. *Mass-Mediated Terrorism*. Oxford: Rowman and Littlefield.

———. 2003. "The Terrorist Calculus behind 9-11: A Model for Future Terrorism?" *Studies in Conflict & Terrorism* 26(1): 1–16.

Nasralla, Shadia. 2013. "Sunni Clerics Call for Jihad Against Assad, Allies." Reuters, June 13. http://www.reuters.com/article/2013/06/13/us-syria-crisis-sunnis-jihad -idUSBRE95C0YQ20130613.

National Commission on Terrorist Attacks upon the United States. 2004. *The 9/11 Commission Report: Final Report of the National Commission on Terrorist Attacks upon the United States.* New York: W. W. Norton.

National Counterterrorism Center. 2012. *2011 Report on Terrorism.* Washington, DC: Office of the Director of National Intelligence, National Counterterrorism Center.

National Research Council. 1991. *Computers at Risk.* Washington, DC: National Academy Press.

Neumann, Peter. 2012. *Countering Online Radicalization in America.* National Security Program Homeland Security Project. December. Washington, DC: Bipartisan Policy Center. http://bipartisanpolicy.org/sites/default/files/BPC%20_Online%20 Radicalization%20Report.pdf.

Nielsen. 2012. *Social Media Report 2012: Social Media Comes of Age.* December 3. http://www.nielsen.com/us/en/insights/news/2012/social-media-report-2012-social -media-comes-of-age.html.

Noel-Levitz. 2012. *E-Expectations Report: The Online Expectations of College-Bound Juniors and Seniors.* https://www.noellevitz.com/documents/shared/Papers_and _Research/2012/2012_E-Expectations.pdf.

Noguchi, Yuki. 2006. "Tracking Terrorists Online." WashingtonPost.com video report transcript. April 19. http://www.washingtonpost.com/wp-dyn/content/discussion /2006/04/11/DI2006041100626.html.

Nye, Joseph. 2003. "The Power of Persuasion: Dual Components of US Leadership (Perspectives on the United States)." Sean Creehan and Sabeel Rahman. *Harvard International Review* 24 (4): 46–49.

———. 2004a. *Power in the Global Information Age: From Realism to Globalization.* London: Routledge.

———. 2004b. "Soft Power and American Foreign Policy." *Political Science Quarterly* 119 (2): 255–70.

Oatley, Keith. 2002. "Emotions and the Story Worlds of Fiction." In *Narrative Impact: Social and Cognitive Foundations,* edited by Melanie C. Green, Jeffrey J. Strange and Timothy C. Brock, 39–69. Mahwah, NJ: Erlbaum.

Obama, Barack. 2014. "Transcript of President Obama's Speech on NSA Reforms." NPR, January 17. http://www.npr.org/blogs/itsallpolitics/2014/01/17/263480199 /transcript-of-president-obamas-speech-on-nsa-reforms.

Office for Security and Counterterrorism. 2009. "The United Kingdom's Strategy for Countering International Terrorism." Home Office of the United Kingdom. http:// www.official-documents.gov.uk/document/cm75/7547/7547.asp.

Oreskovic, Alexei. 2013. "Facebook Removes Beheading Video, Updates Violent Images Standards." Reuters UK, October 23. http://uk.reuters.com/article/2013/10 /23/uk-facebook-violence-idUKBRE99M01O20131023.

Pantucci, Raffaello. 2011. "Typology of Lone Wolves: Preliminary Analysis of Lone Islamist Terrorists." *Developments in Radicalisation and Political Violence,* March. London: The International Centre for the Study of Radicalisation and Political Violence. http://www.trackingterrorism.org/sites/default/files/chatter /1302002992ICSRPaper_ATypologyofLoneWolves_Pantucci.pdf.

Paz, Reuven. 2003. "Sawt al-Jihad: New Indoctrination of Qa'idat al-Jihad." Project for the Study of Islamist Movements (PRISM) Occasional Paper 1, No. 8. http://www.e-prism.org/images/PRISM_no_8.doc.

———. 2004. "Hamas vs. Al-Qaeda: The Condemnation of the Khobar Attack." Project for the Study of Islamist Movements (PRISM) Special Dispatch 2 (3), June 2. http://www.e-prism.org/images/PRISM_Special_dispatch_no_3-2.pdf.

———. 2005. "Global Jihad and WMD: Between Martyrdom and Mass Destruction." *Current Trends in Islamist Ideology* 2: 74–86.

———. 2010. "Global Jihad." In *Guide to Islamist Movements*, Vol. 2, edited by Barry M. Rubin, xxxiii–liii. Armonk, NY: M. E. Sharpe.

PearAnalytics. 2009. *PearAnalytics: Twitter Study—August 2009*. August. https://www.pearanalytics.com/wp-content/uploads/2012/12/Twitter-Study-August-2009.pdf.

Perl, Raphael. 2004. *The Department of State's Patterns of Global Terrorism Report: Trends, State Sponsors, and Related Issues*. Congressional Research Service Report for Congress RL32417, June 5. Washington, DC: Congressional Research Service.

Perlroth, Nicole, and David E. Sanger. 2013. "Cyberattacks Seem Meant to Destroy, Not Just Disrupt." *New York Times*, March 28. http://www.nytimes.com/2013/03/29/technology/corporate-cyberattackers-possibly-state-backed-now-seek-to-destroy-data.html.

Pew Research Center. 2007. "Muslim Americans: Middle Class and Mostly Mainstream." May 22. http://www.pewresearch.org/2007/05/22/muslim-americans-middle-class-and-mostly-mainstream/.

Pew Research Center for the People and the Press. 2013. "Majority Views NSA Phone Tracking as Acceptable Anti-Terror Tactic." Pew Research Center/*Washington Post* Survey, June 10. http://www.people-press.org/2013/06/10/majority-views-nsa-phone-tracking-as-acceptable-anti-terror-tactic/.

Picali, E. B. 2013a. "Independent Shi'ites in Lebanon Challenge Hizbullah." MEMRI Inquiry & Analysis Series Report 983, February 22. http://www.memri.org/report/en/0/0/0/0/259/0/7017.htm.

———. 2013b. "Rift in Hizbullah and among Its Shi'ite Supporters Due to Its Military Involvement in Syria." MEMRI Inquiry & Analysis Series Report 1021, October 3. http://www.memri.org/report/en/0/0/0/0/259/0/7438.htm.

Pollack, Suzanne. 2013. "Internet provides venue for training future jihadists." *Washington Jewish Week*, May 16: 3.

Porter, Michael E. 2001. "Strategy and the Internet." *Harvard Business Review*, March: 62–78.

President's Commission on Critical Infrastructure Protection. 1997. *Critical Foundations: Protecting America's Infrastructures*. October. Project on Government Secrecy, Federation of American Scientists. Accessed April 3, 2014. https://www.fas.org/sgp/library/pccip.pdf.

Presidential Task Force. 2009. *Rewriting the Narrative: An Integrated Strategy for Counterradicalization*. March. Washington, DC: Washington Institute for Near East Policy. https://www.washingtoninstitute.org/uploads/Documents/pubs/PTF2-Counterradicalization.pdf.

Priest, Dana, and William M. Arkin. 2010. "Hidden World, Growing Beyond Control." *Washington Post*, July 19. http://projects.washingtonpost.com/top-secret-america/articles/a-hidden-world-growing-beyond-control/.

Project on Muslims in the American Public Square (Project MAPS). 2002. "American Muslim Poll." Georgetown University. May 2002. Accessed January 4, 2014. http://www.projectmaps.com/PMReport.htm.

Prucha, Nico, and Ali Fisher. 2013. "Tweeting for the Caliphate: Twitter as the New Frontier for Jihadist Propaganda." *CTC Sentinel* 6 (6): 19–22. https://www.ctc.usma.edu/posts/tweeting-for-the-caliphate-twitter-as-the-new-frontier-for-jihadist-propaganda.

Qaradawi, Sheikh Yusuf al-. 2003. "Fatwa: Whether Suicide Bombings Are a Form of Martyrdom" [in Arabic]. November 8. http://qaradawi.net.

Qin, Jialun, Yilu Zhou, Edna Reid, Guanpi Lai, and Hsinchun Chen. 2007. "Analyzing Terror Campaigns on the Internet: Technical Sophistication, Content Richness, and Web Interactivity." *International Journal of Human-Computer Studies* 65 (1): 71–84.

Ranstorp, Magnus. 1998. "Interpreting the Broader Context and Meaning of Bin-Laden's 'Fatwa.'" *Studies in Conflict & Terrorism* 21 (4): 321–30.

———. 2007. "The Virtual Sanctuary of Al-Qaeda and Terrorism in an Age of Globalization." In *International Relations and Security in the Digital Age*, edited by Johan Eriksson and Giampiero Giacomello, 31–56. London: Routledge.

Reform Government Surveillance. 2013. "Global Government Surveillance Reform." Accessed March 12, 2014. https://www.reformgovernmentsurveillance.com/#.

Risen, James, and Eric Lichtblau. 2005. "Bush Lets U.S. Spy on Callers Without Courts." *New York Times*, December 16. http://www.nytimes.com/2005/12/16/politics/16program.html.

———. 2009. "Officials Say U.S. Wiretaps Exceeded Law." *New York Times*, April 15. http://www.nytimes.com/2009/04/16/us/16nsa.html.

Rogan, Hanna. 2006. *Jihadism Online: A Study of How Al-Qaida and Radical Islamist Groups Use the Internet for Terrorist Purposes*. Norwegian Defense Research Establishment. http://rapporter.ffi.no/rapporter/2006/00915.pdf.

Rollins, John, and Clay Wilson. 2007. *Terrorist Capabilities for Cyberattack: Overview and Policy Issues*. Congressional Research Service Report for Congress RL 33123, January 22. Washington, DC: Congressional Research Service.

Rothman, Paul. 2012. "Cyber Terror Rages in the Banking Sector." *Security InfoWatch* 28, September. http://www.securityinfowatch.com/blog/10796084/cyber-terror-rages-in-the-banking-sector.

Rothwell, Dan. 2004. *In the Company of Others: An Introduction to Communication*. New York: McGraw Hill.

Roy, Olivier. 2008a. "Al-Qaeda in the West as a Youth Movement: The Power of a Narrative." MicroCon Policy Working Paper 2. http://www.microconflict.eu/publications/PWP2_OR.pdf

———. 2008b. "Radicalisation and De-Radicalisation." In *Perspectives on Radicalisation and Political Violence*, edited by the International Centre for the Study of Radicalisation (ICSR), 8–14. London: ICSR.

Rumsfeld, Donald. 2006. "New Realities in the Media Age." Speech at the Council on Foreign Relations, February 17. http://www.cfr.org/publication/9900/new_realities_in_the_media_age.html.

Sageman, Marc. 2008. *Leaderless Jihad: Terror Networks in the Twenty-First Century*. Philadelphia: University of Pennsylvania Press.

Sammy67 [pseud]. 2008. "Al-Qaida Eyes U.S. Troops MySpace Pages." FreeRepublic.com, January 14. http://www.freerepublic.com/focus/f-news/1953538/posts.

Sanger, David E., and John O'Neil. 2006. "White House Begins New Effort to Defend Surveillance Program." *New York Times*, January 23. http://www.nytimes.com/2006 /01/23/politics/23cnd-wiretap.html.

Satter, Raphael. 2013. "NSA Hacking Tactics Revealed By Der Spiegel." *Huffington Post*, December 29: 2–13. http://www.huffingtonpost.com/2013/12/29/nsa-hacking -tactics-_n_4515897.html.

"SCADA Systems and the Terrorist Threat: Protecting the Nation's Critical Control Systems." 2005. Joint hearing before the US House of Representatives Subcommittee on Economic Security, Infrastructure Protection and Cybersecurity, October 18. Intelligence Resource Program, Federation of American Scientists. Accessed April 10, 2014. http://www.fas.org/irp/congress/2005_hr/scada.pdf.

Schmid, Alex. 2004. "Frameworks for Conceptualising Terrorism." *Terrorism and Political Violence* 16 (2): 197–221.

———. 2010. "The Importance of Countering Al-Qaeda's 'Single Narrative.'" In *Countering Violent Extremist Narratives*, edited by Eelco Kessels, 46–57. The Hague: The National Coordinator for Counter-Terrorism (NCTb). http://www .clingendael.nl/sites/default/files/Countering-violent-extremist-narratives.pdf.

Schmid, Alex, and Janny de Graaf. 1982. *Violence as Communication: Insurgent Terrorism and the Western News Media*. Beverly Hills, CA: Sage.

Schmid, Alex, and Albert Jongman. 1988. *Political Terrorism*. Amsterdam: Transaction Books.

———. 2005. *Political Terrorism*. Piscataway, NJ: Transaction Publishers.

Schmidt, Eric, and Jared Cohen. 2013. *The New Digital Age: Reshaping the Future of People, Nations and Business*. New York: Knopf Doubleday Publishing Group.

Schmidt, Michael S. 2012. "F.B.I. Director Warns Congress About Terrorist Hacking." *New York Times*, March 7. http://www.nytimes.com/2012/03/08/us/fbi-director -warns-about-terrorist-hacking.html.

Schmitt, Eric, and Michael S. Schmidt. 2013. "Qaeda Plot Leak Has Undermined U.S. Intelligence." *New York Times*, September 29. http://www.nytimes.com/2013/09/30 /us/qaeda-plot-leak-has-undermined-us-intelligence.html.

Schmitt, Eric, and Thom Shanker. 2011. *Counterstrike: The Untold Story of America's Secret Campaign Against Al Qaeda*. New York: Times Books.

Schneier, Bruce. 2009. "Terrorists May Use Google Earth, but Fear Is No Reason to Ban It." *Guardian* (London), January 28. http://www.theguardian.com/technology /2009/jan/29/read-me-first-google-earth.

Schwartz, John. 2001. "Tools for the Aftermath: In Investigation, Internet Offers Clues and Static." *New York Times*, September 26: H1. http://www.nytimes.com/2001/09 /26/business/tools-for-the-aftermath-in-investigation-internet-offers-clues-and -static.html.

Shane, Scott. 2013. "A Homemade Style of Terror: Jihadists Push New Tactics." *New York Times*, May 5. http://www.nytimes.com/2013/05/06/us/terrorists-find-online -education-for-attacks.html.

Sherif, Muzafer, and Carl Iver Hovland. 1961. *Social Judgment: Assimilation and Contrast Effects in Communication and Attitude Change*. New Haven, CT: Yale University Press.

Shiffman, John. 2010. "Jane's Jihad: The New Face of Terrorism." Reuters, March 9. http://www.reuters.com/subjects/jihad-jane.

Shih, Gerry, and Joseph Menn. 2013. "New York Times, Twitter Hacked by Syrian Group." Reuters, August 28. http://www.reuters.com/article/2013/08/28/media -hacking-idUSL2N0GS29C20130828.

Siddiqui, Sabrina, and Jaweed Kaleem. 2013. "Muslims Focus on Online Extremism, Radicalization after Boston Bombings." *Huffington Post*, June 4. http://www .huffingtonpost.com/2013/06/04/muslims-online-extremism-radicalization-boston _n_3380159.html.

Sinai, Joshua. 2006. "Defeating Internet Terrorists." *Washington Times*, October 7. http://www.washingtontimes.com/news/2006/oct/7/20061007-104915-3656r/.

——. 2011. "Terrorism on the Internet and Effective Countermeasures." *The Intelligencer: Journal of U.S. Intelligence Studies* 18: 21–24.

SITE Monitoring Service. 2005. "Salafi Group for Call and Combat Issues Fatwa Calling for Jihad Against Foreigners in Algeria." March 11. http://ent.siteintelgroup .com/Jihadist-News/salafi-group-for-call-and-combat-issues-fatwa-calling-for-jihad -against-foreigners-in-algeria.html.

——. 2008. "Al-Nusra Media Battalion Distributes Guide for 'Martyrdom.'" December 18. http://ent.siteintelgroup.com/Jihadist-News/al-nusra-media-battalion -distributes-guide-for-martyrdom.html.

——. 2011a. "Jihadist Offers Tips for Facebook Users to Increase Page Viewership." June 2. https://news.siteintelgroup.com/index.php/21-social-network-jihad/776 -jihadist-offers-tips-for-facebook-users-to-increase-page-viewership.

——. 2011b. "Jihadists Strategize to Evade YouTube Censorship." April 28. http:// ent.siteintelgroup.com/Social-Network-Jihad/site-intel-group-4-28-11-jfm-youtube -strategies.html.

——. 2011c. "Jose Pimentel and the Use of Social Networks for Jihadist Recruitment." January 14. https://news.siteintelgroup.com/Social-Network-Jihad/jose-pimentel-and -the-use-of-social-networks-for-jihadist-recruitment.html.

——. 2012a. "'Cyber Fighters' Announces 'Phase 2' of Banking Website Hacks." December 11. http://ent.siteintelgroup.com/Jihadist-News/cyber-fighters-announces -phase-2-of-banking-website-hacks.html.

——. 2012b. "Jihadist Gives Analysis of Electronic Jihad." January 6. http://news. siteintelgroup.com/index.php/19-jihadist-news/1462-jihadist-gives-analysis-of -electronic-jihad.

——. 2013a. "'Al-Qaeda Electronic Army' Threatens to Hit Vital Sectors of the US." April 12. http://ent.siteintelgroup.com/Jihadist-News/al-qaeda-electronic-army -threatens-to-hit-vital-sectors-of-us.html.

——. 2013b. "AQIM Blames Hezbollah for Tripoli Bombings, Promises Retribution." August 23. http://ent.siteintelgroup.com/Jihadist-News/aqim-blames-hezbollah-for -tripoli-bombings-promises-retribution.html.

——. 2013c. "Facebook Page Serves as Official Facebook Outlet for Ansar al-Islam." August 6. https://news.siteintelgroup.com/index.php/19-jihadist-news/3342-facebook -page-serves-as-official-facebook-outlet-for-ansar-al-islam.

——. 2013d. "Jihadists Create Twitter-based Jihadi Media Group, Al-Battar Media Battalion." July 19. https://news.siteintelgroup.com/Jihadist-News/jihadists-create -twitter-based-jihadi-media-group-al-battar-media-battalion.html.

——. 2014a. "Female Jihadists Promote Attacks in West in Third Issue of Magazine." January 15. https://news.siteintelgroup.com/Jihadist-News/female-jihadists-promote -attacks-in-west-in-third-issue-of-magazine.html.

————. 2014b. "Jihadist Invites to 'Electronic Islamic Army,' Gives DDOS Program Tutorial." January 15. https://news.siteintelgroup.com/Jihadist-News/jihadist-invites -to-qelectronic-islamic-armyq-gives-ddos-program-tutorial.html.

————. 2014c. "Social Network Jihad: Hezbollah's Capitalization of Facebook." January 14. https://news.siteintelgroup.com/Social-Network-Jihad/social-network -jihad-hezbollahs-capitalization-of-facebook.html.

————. 2014d. "The Third Palestinian Intifada's Facebook Page." January 15. https:// news.siteintelgroup.com/Featured-Article/the-third-palestinian-intifadas-facebook -page.html.

Slater, Michael. 2002. "Entertainment Education and the Persuasive Impact of Narratives." In *Narrative Impact: Social and Cognitive Foundations*, edited by Melanie C. Green, Jeffrey J. Strange, and Timothy C. Brock, 157–81. Mahwah, NJ: Erlbaum.

Smith, Paul Russell, and Dave Chaffey. 2001. *eMarketing Excellence: At the Heart of eBusiness*. Oxford: Butterworth Heinemann.

The Soufan Group. 2013. *TSG IntelBrief: The Importance of Counter Narratives in Fighting Violent Extremism*. September 10. http://soufangroup.com/tsg-intelbrief -3/.

Spaaij, Ramon. 2010. "The Enigma of Lone Wolf Terrorism: An Assessment." *Studies in Conflict & Terrorism* 33 (9): 854–70.

Spring, Tom. 2004. "Al Qaeda's Tech Traps." *PCWorld*, September 1. http://www .pcworld.com/article/117658/article.html.

Stalinsky, Steven. 2012a. "HASHTAG #*Jihad* Part II: Twitter Usage By Al-Qaeda And Online Jihadi Affiliated Groups Explodes." MEMRI Inquiry & Analysis Series Report 881, September 7. http://www.memri.org/report/en/0/0/0/0/0/0/6660.htm.

————. 2012b. "Tracking Hizbullah's Al-Manar TV Online: Removed from Servers in Netherlands and U.S. as a Result of MEMRI Reports—and Now Hosted on Servers in U.K." MEMRI Inquiry & Analysis Series Report 905, December 4. http://www .memri.org/report/en/0/0/0/0/857/6848.htm

————. 2012c. "WHOIS—U.S.-Based Domain Registration Protection Companies Hide Identities Of Individuals Behind The Most Important Al-Qaeda-Affiliated Websites." MEMRI Inquiry & Analysis Series Report 856, July. http://www.memri .org/report/en/0/0/0/0/0/0/6515.htm.

————. 2013a. "Hizbullah—A Victim of Identity Theft." MEMRI Inquiry & Analysis Series Report 985, June 20. http://www.memri.org/report/en/0/0/0/0/0/0/7252.htm.

————. 2013b. "Jihadists Embrace Instagram." MEMRI Inquiry & Analysis Series Report 468, March 14. http://www.memri.org/report/en/0/0/0/0/0/857/7081.htm.

————. 2013c. "Jihadis Move From Facebook To Yahoo's Flickr Picture Sharing Site." MEMRI Inquiry & Analysis Series Report 1005, August 5. http://www.memri.org /report/en/0/0/0/0/0/857/7334.htm.

Stalinsky, Steven, and R. Sosnow. 2013a. "Afghan Cyber Army (ACA): Active on Facebook and YouTube—Recruiting and Training Volunteers, Claiming to Have Hacked U.S. Gov't Websites, U.S. and Israeli Bank Websites, Pakistan Gov't and Taliban Websites; Promising Major Attacks in Future." MEMRI Inquiry & Analysis Series Report 1036, November 15. http://www.memri.org/report/en/0/0/0/0/0/0/7575 .htm.

————. 2013b. "Syrian Electronic Army Uses Social Media—Twitter, YouTube, Facebook, Instagram, Google+, Pinterest, Smartphone Apps—To Communicate,

Spread News of Its Hacks and Its Mission, and Recruit Volunteers." MEMRI Inquiry & Analysis Series Report 1009, August 19. http://www.memri.org/report/en/0/0/0/0/0/0/7357.htm.

Stalinsky, Steven, and Elliot Zweig. 2013. "YouTube Questioned in U.K. House of Commons over Keeping Terrorism-Promoting Videos Active on Its Website." MEMRI Inquiry & Analysis Series Report 956, April 9. http://www.memri.org/report/en/0/0/0/0/0/0/841/7121.htm.

Statistic Brain. 2013. "Twitter Statistics." Statistic Brain, May 7. Accessed March 12, 2014. http://www.statisticbrain.com/twitter-statistics/.

Stevens, Tim, and Peter Neumann. 2009. *Countering Online Radicalization: A Strategy for Action*. London: International Centre for the Study of Radicalisation and Political Violence (ICSR). http://www.thecst.org.uk/docs/countering_online_radicalisation1.pdf.

Stevenson, Robert. 2003. "Freedom of the Press Around the World." In *Global Journalism: Topical Issues and Media Systems*, 4th edition, edited by Arnold S. de Beer and John C. Merrill, 1–33. Boston: Allyn and Bacon.

Strohm, Chris, and Del Quentin Wilber. 2014. "Pentagon Says Snowden Took Most U.S. Secrets Ever: Rogers." Bloomberg.com, January 9. http://www.bloomberg.com/news/2014-01-09/pentagon-finds-snowden-took-1-7-million-files-rogers-says.html.

Sullivan, Bob. 2001. "FBI Software Cracks Encryption Wall." NBCNews.com, November 20. http://www.nbcnews.com/id/3341694/ns/technology_and_science-security/t/fbi-software-cracks-encryption-wall/.

Swartz, Jon. 2005. "Terrorists' Use of Internet Spreads." *USA Today*, February 20. http://www.usatoday.com/money/industries/technology/2005-02-20-cyber-terror-usat_x.htm.

"Syria: Speech by Bashar al-Assad." 2011. Al-Bab.com, June 20. http://www.al-bab.com/arab/docs/syria/bashar_assad_speech_110620.htm.

Syrian Electronic Army. 2014. "About Organization." Accessed September 30, 2014. http://sea.sy/about/en.

Tahawy, Abdallah, el-. 2008. "The Internet Is the New Mosque: Fatwa at the Click of a Mouse." *Arab Insight* 2 (1):11–20.

Talbot, David. 2005. "Terror's Server." *Technology Review*, February 1. http://www.technologyreview.com/featuredstory/403657/terrors-server/.

Teich, Sara. 2013. *Trends and Developments in Lone Wolf Terrorism in the Western World: An Analysis of Terrorist Attacks and Attempted Attacks by Islamic Extremists*. Herzliya, Israel: International Institute for Counter-Terrorism. http://www.ict.org.il/Article/691/Trends%20and%20Developments%20in%20Lone%20Wolf%20Terrorism%20in%20the%20Western%20World.

Tell, William. 2006. *Bridges Burning: America's Challenge of the 21st Century*. Lincoln, NE: iUniverse.

Thomas, Timothy. 2003. "Al Qaeda and the Internet: The Danger of 'Cyberplanning.'" *Parameters* (Spring): 112–23. http://strategicstudiesinstitute.army.mil/pubs/parameters/articles/03spring/thomas.pdf.

Tsfati, Yariv, and Gabriel Weimann. 2002. "www.terrorism.com: Terror on the Internet." *Studies in Conflict & Terrorism* 25 (5): 317–32.

United Nations Counter-Terrorism Implementation Task Force. 2011. *Countering the Use of the Internet for Terrorist Purposes: Legal and Technical Aspects*. New York: United Nations.

———. 2012. "Use of the Internet to Counter the Appeal of Extremist Violence. Conference Summary & Follow-up/Recommendations." *Perspectives on Terrorism* 6 (1): 80–91. http://www.terrorismanalysts.com/pt/index.php/pot/article/view /CTITF-Use-of-Internet/html.

United Nations Office on Drugs and Crime (UNODC). 2012. *The Use of the Internet for Terrorist Purposes.* New York: United Nations.

United Nations Security Council. 2009. "Security Council Committee pursuant to resolutions 1267 (1999) and 1989 (2011) concerning Al-Qaida and associated individuals and entities." Accessed October 13, 2013. http://www.un.org/sc /committees/1267/NSQI23608E.shtml.

———. 2014. "Al-Qaida Sanctions List." Accessed October 1, 2014. http://www.un.org /sc/committees/1267/AQList.htm.

Vaccani, Matteo. 2010. *Alternative Remittance Systems and Terrorism Financing.* Working Paper No. 1980. First printed November 2009. Washington, DC: World Bank. http://elibrary.worldbank.org/doi/pdf/10.1596/978-0-8213-8178-6.

Vatis, Michael. 2001. *Cyber Attacks during the War on Terrorism: A Predictive Analysis.* Institute for Security Technology Studies, Dartmouth College. http://www.ists .dartmouth.edu/docs/cyber_a1.pdf.

Ventre, Daniel. 2009. *Information Warfare.* New York: Wiley.

———, ed. 2011. *Cyberwar and Information Warfare.* New York: Wiley.

Verton, Dan. 2003. *Black Ice: The Invisible Threat of Cyber-Terrorism.* New York: McGraw-Hill Osborne Media.

Viscusi, W. Kip, and Richard J. Zeckhauser. 2003. "Sacrificing Civil Liberties to Reduce Terrorism Risks." *Journal of Risk and Uncertainty* 26 (2–3): 99–120.

Von Knop, Katharina, and Gabriel Weimann. 2008. "Applying the Notion of Noise to Countering Online Terrorism." *Studies in Conflict & Terrorism* 31 (10): 883–902.

Voors, Matthew Parker. 2003. "Encryption Regulation in the Wake of September 11, 2001: Must We Protect National Security at the Expense of the Economy?" *Federal Communications Law Journal* 55 (2): 331–52.

Wade, Lindsey. 2003. "Terrorism and the Internet: Resistance in the Information Age." *Knowledge, Technology & Policy* 16 (1): 104–27.

Weaver, Warren, and Claude Shannon. 1963. *The Mathematical Theory of Communication.* Urbana: University of Illinois Press.

Weimann, Gabriel. 2005a. "Cyberterrorism: The Sum of All Fears?" *Studies in Conflict & Terrorism* 28 (2): 129–49.

———. 2005b. "How Terrorists Use the Internet." *Journal of International Security Affairs* 8: 91–105.

———. 2005c. "Terrorist Dot Com: Using the Internet for Terrorist Recruitment and Mobilization." In *The Making of a Terrorist: Recruitment, Training, and Root Causes,* edited by James J. F. Forest, 53–65. Westport, CT: Praeger.

———. 2006a. "Cyberterrorism." In *Security, Terrorism and Privacy in Information Society,* edited by Katharina Von Knop and Boaz Ganor, 41–52. Bielefeld, Germany: W. Bertelsmann Verlag.

———. 2006b. *Terror on the Internet: The New Arena. The New Challenges.* Washington, DC: United States Institute for Peace (USIP) Press.

———. 2006c. "Virtual Disputes: The Use of the Internet for Terrorist Debates." *Studies in Conflict & Terrorism* 29 (7): 623–39.

———. 2006d. "Virtual Training Camps: Terrorist Use of the Internet." In *Teaching Terror: Strategic and Tactical Learning in the Terrorist World*, edited by James Forest, 110–32. Boulder, CO: Rowman & Littlefield.

———. 2007a. "Using the Internet for Terrorist Recruitment and Mobilization." In *Hypermedia Seduction for Terrorist Recruiting*, edited by Boaz Ganor, Katharina Von Knop, and Carlos Duarte, 47–58. NATO Science for Peace and Security Series. Amsterdam: IOS Press.

———. 2007b. "Virtual Terrorism: How Modern Terrorists Use the Internet." In *The Internet and Governance in Asia: A Critical Reader*, edited by Indrajit Banerjee, 189–216. Singapore: Asian Media Information and Communication Centre and Wee Kim Wee School of Communication and Information, Nanyang Technological University.

———. 2008a. "Al-Qa'ida's Extensive Use of the Internet." *CTC Sentinel*. January 15. http://www.ctc.usma.edu/posts/al-qaida%E2%80%99s-extensive-use-of-the -internet.

———. 2008b. "Cyber-Terrorism: Are We Barking at the Wrong Tree?" *Harvard Asia Pacific Review* 9 (2) (Spring 2008): 41–46.

———. 2008c. "How Terrorists Use the Internet to Target Children." *InSite* 1 (8): 14–16. http://sitemultimedia.org/docs/inSITE_December_2008.pdf.

———. 2008d. "Online Terrorists Prey on the Vulnerable." *Yale Global Online*. March 5. http://yaleglobal.yale.edu/content/online-terrorists-prey-vulnerable.

———. 2008e. "The Psychology of Mass-Mediated Terrorism." *American Behavioral Scientist* 52 (1): 69–86.

———. 2008f. "WWW.Al-Qaeda: The Reliance of al-Qaeda on the Internet." In *Responses to Cyber Terrorism*, edited by Centre of Excellence—Defence Against Terrorism, Ankara, Turkey, 61–69. NATO Science for Peace and Security Series. Amsterdam: IOS Press.

———. 2009a. "Online Training Camps for Terrorists," *InSite*, Vol. 2 No. 9. http:// sitemultimedia.org/docs/inSITE_Nov_2009.pdf.

———. 2009b. "Virtual Sisters: How Terrorists Target Women Online." *InSite* 2 (1): 19–22. http://sitemultimedia.org/docs/inSITE_January_2009.pdf.

———. 2009c. "War by Other Means: Econo-Jihad." *Yale Global Online*. June 4. http://yaleglobal.yale.edu/content/econo-jihad.

———. 2009d. "When Fatwas Clash Online: Terrorist Debates on the Internet." In *Influence Warfare: How Terrorists and Governments Fight to Shape Perceptions in a War of Ideas*, edited by James Forest, 49–74. Westport, CT: Praeger Security International.

———. 2010a. "Terror on Facebook, Twitter, and Youtube." *The Brown Journal of World Affairs* 16 (2): 45–54. http://brown.edu/initiatives/journal-world-affairs/16.2 /terror-facebook-twitter-and-youtube.

———. 2010b. "Terrorism's New Avatars—Part II." *Yale Global Online*. January 12. http://yaleglobal.yale.edu/content/terrrorisms-new-avatars-part-ii.

———. 2010c. "Terrorist Facebook: Terrorists and Online Social Networking." In *Web Intelligence and Security*, edited by Mark Last and Abraham Kandel, 19–30. NATO Science for Peace and Security Series. Amsterdam: IOS Press.

———. 2011a. "Al Qaeda Has Sent You a Friend Request: Terrorists Using Online Social Networking." Paper presented at annual conference of the Israeli Communication Association, Haifa, Israel, April 14.

————. 2011b. "Cyber-*Fatwas* and Terrorism." *Studies in Conflict & Terrorism* 34 (10): 765–81.

————. 2012a. "Lone Wolves in Cyberspace." *Journal of Terrorism Research* 3 (2): 75–90.

————. 2012b. "The Role of the Media in Propagating Terrorism." In *Countering Terrorism: Psychosocial Strategies*, edited by Updesh Kumar and Manas K. Mandal, 182–200. London: Sage Publications.

————. 2014a. *New Terrorism and New Media*. Commons Lab, Science and Technology Innovation Program. Washington, DC: Woodrow Wilson International Center for Scholars. http://www.wilsoncenter.org/publication/new-terrorism-and-new-media.

————. 2014b. "Virtual Packs of Lone Wolves." Medium.com/@thewilsoncenter, February 28. https://medium.com/its-a-medium-world/virtual-packs-of-lone-wolves-17b12f8c455a.

Weimann, Gabriel, and Gabrielle Vail Gorder. 2009. "Al-Qaeda Has Sent You A Friend Request: Terrorists Using Online Social Networking." *InSite* 2:6. http://sitemultimedia.org/docs/inSITE_June_2009.pdf.

Weimann, Gabriel, and Conrad Winn. 1994. *The Theater of Terror: Mass Media and International Terrorism*. New York: Longman.

Westby, Jody. 2006. "Countering Terrorism with Cyber Security." Paper presented at the 36th Session of World Federation of Scientists, International Seminars on Planetary Emergencies, August 18–26, 2006, Erice, Italy.

Wheeler, Ashley. 2013. "Iranian Cyber Army, the Offensive Arm of Iran's Cyber Force." Phoenix TS blog, September 19. http://www.phoenixts.com/blog/iranian-cyber-army/#sthash.qMRqtrQg.1Pl02y9b.dpuf.

The White House. 2011a. *Empowering Local Partners to Prevent Violent Extremism in the United States*. August 2011. Washington, DC: The White House. http://www.whitehouse.gov/sites/default/files/empowering_local_partners.pdf.

————. 2011b. *Strategic Implementation Plan for Empowering Local Partners to Prevent Violent Extremism in the United States*. December 2011. Washington, DC: The White House. http://www.whitehouse.gov/sites/default/files/sip-final.pdf.

Wilkinson, Paul. 2001. *Terrorism versus Democracy*. London: Frank Cass.

Wilson, Clay. 2005. *Computer Attack and Cyberterrorism: Vulnerabilities and Policy Issues for Congress*. Congressional Research Service Report for Congress RL32114, April 1. Washington, DC: Congressional Research Service.

————. 2007. *Information Operations, Electronic Warfare, and Cyberwar: Capabilities and Related Policy Issues*. Congressional Research Service Report for Congress RL31787, June 5. Washington, DC: Congressional Research Service.

————. 2008. *Botnets, Cybercrime, and Cyberterrorism: Vulnerabilities and Policy Issues for Congress*. Congressional Research Service Report for Congress RL32114, January 29. Washington, DC: Congressional Research Service.

Withnall, Adam. 2014. "Iraq Crisis: ISIS Declares Its Territories a New Islamic State with 'Restoration of Caliphate' in Middle East." *The Independent*, June 30. http://www.independent.co.uk/news/world/middle-east/isis-declares-new-islamic-state-in-middle-east-with-abu-bakr-albaghdadi-as-emir-removing-iraq-and-syria-from-its-name-9571374.html.

Wolfsfeld, Gadi, Eli Avraham, and Issam Aburaiya. 2001. "When Prophesy Always Fails: Israeli Press Coverage of the Arab Minority's Land Day Protests." *Political Communication* 17 (2): 115–31.

World Public Opinion. 2009. *Public Opinion in the Islamic World on Terrorism, al Qaeda, and US Policies*. February 25. WorldPublicOpinion.org. Program on International Policy Attitudes, University of Maryland. http://www .worldpublicopinion.org/pipa/pdf/feb09/STARTII_Feb09_rpt.pdf.

Wright, Lawrence. 2007. *The Looming Tower: Al-Qaeda and the Road to 9/11*. New York: Alfred A. Knopf.

Youssef, Maamoun, and Lee Keath. 2014. "Al-Qaida Breaks with Syria Group in Mounting Feud." Associated Press, February 3. http://bigstory.ap.org/article/al-qaida -breaks-syria-group-mounting-feud.

YouTube. 2014. "Statistics." Accessed March 12, 2014. http://www.youtube.com/yt/press /statistics.html.

Zanini, Michele, and Sean J. A. Edwards. 2001. "The Networking of Terror in the Information Age." In *Networks and Netwars: The Future of Terror, Crime and Militancy*, edited by John Arquilla and David Ronfeldt, 29–60. Santa Monica, CA: RAND Corporation.

Zanna, Mark P. 1993. "Message Receptivity: A New Look at the Old Problem of Open - vs. Closedmindedness." In *Advertising Exposure, Memory and Choice*, edited by Andrew A. Mitchell, 141–62. Hillsdale, NJ: Erlbaum.

Zelin, Aaron. 2013. *The State of Global Jihad Online*. January. Washington, DC: New America Foundation. http://www.washingtoninstitute.org/uploads/Documents /opeds/Zelin20130201-NewAmericaFoundation.pdf.

Zetter, Kim. 2011. "DHS Fears a Modified Stuxnet Could Attack U.S. Infrastructure." Wired.com, July 26. http://www.wired.com/threatlevel/2011/07/dhs-fears-stuxnet -attacks/.

Index